A1
LANDMARKS
FASCINATING POINTS ALONG THE ROUTE

JAMES CLARK

Acknowledgements

Peter Fisher at London Metropolitan University (entry 16)
Dylan Mills (entry 28)
Ferrers Young at the BWTAS (entry 29)
David Hinds at the Sibson Inn (entry 34)
Keith Clark at CBP Architects (entry 41)
Zulumike and Lesc on pylons.proboards.com (entry 39)
Alistair Snart at RMS (Darrington) Ltd (entry 50)
Katy Harris at Foster & Partners (entry 81)
Roy Smith
Elain Harwood
Jackie Clark
Anita Clark

Images
All images © James Clark with the exception of:

Entry 32 © Richard White
Entry 37 © Kevin Pye
Entry 65 © Joe Bright

First published 2015

Amberley Publishing
The Hill, Stroud
Gloucestershire, GL5 4EP

www.amberley-books.com

Copyright © James Clark, 2015

The right of James Clark to be identified as the
Author of this work has been asserted in accordance
with the Copyrights, Designs and Patents Act 1988.

ISBN 978 1 4456 5450 8 (print)
ISBN 978 1 4456 5451 5 (ebook)

British Library Cataloguing in Publication Data.
A catalogue record for this book is available from
the British Library.

Typeset in 9.5pt on 12pt Celeste.
Typesetting by Amberley Publishing.
Printed in the UK.

Contents

Introduction

In truth, it isn't the most inspiring location for the beginning (or the end) of an epic journey. A roundabout formed where the western end of the London Wall urban dual carriageway meets Aldersgate Street marks the point where Britain's longest numbered road begins its long journey north. There are no visual clues that this is mile zero of the A1, no 'Edinburgh 396' signpost or tourist trail information board. Instead there are only the blank faces of office blocks peering down on the Museum of London's garden rotunda. The A1 will perhaps never inspire the same level of affection as some of our other long-distance routes such as the Pennine Way or the Grand Union Canal; for most people it is too closely connected with the daily drudgery of commuting, traffic jams and roadworks. They may traverse the same section hundreds of time a year, passing the same landmarks without ever stopping or reflecting on their history. For as the social historian Joe Moran has noted, the roadside is 'the most commonly viewed and least contemplated landscape in Britain' (Joe Moran, *On Roads – A Hidden History* (page 147)).

This indifference is a pity, since the built environment in the immediate vicinity of the A1 contains hundreds of interesting structures that should intrigue any keen student of architecture, engineering or the visual arts. An entire millennium's worth of development in these fields can be discerned. There are fragments of richly decorated Romanesque churches in sleepy rural backwaters that contrast with 1960s modernist-influenced places of worship in busy urban settings. We can stumble across relics from the Industrial Revolution: abandoned coal mines and limekilns that are dwarfed by the monolithic bulk of modern nuclear or coal-fired power stations. Traditional corn mills and agricultural windpumps have given way to serried ranks of high-tech wind turbines. There are half-timbered cottages, their structures distorted by age and shaken by passing articulated lorries, and there are sleek residential high-rises casting long shadows across their inner-city neighbourhoods.

The A1 has sometimes been compared, usually in a slightly disparaging tone, to Route 66. While it's certainly true that it has not been mythologised in popular culture in the same way as the Mother Road of America, our own long-distance highway can justifiably claim to be at least an equal in terms of architectural treasures. Many of these can be found at the northern and southernmost extremities of the route, as the road snakes its way into the heart of two historic capital cities. In between London and Edinburgh the modern A1 is largely composed of fast-running dual carriageway and motorway sections with any towns and cities having long since been bypassed. Yet there is still much to see. This book will attempt to shed some light on the past, present and future of some of these landmarks. In several cases their description here might serve as some kind of record before they are irrevocably altered or demolished. For this is a road in a constant state of flux, with relentless changes to the route and its environs. As the A1 approaches its 100th birthday it is time to celebrate this uniquely British highway and the diverse structures that line its path.

Criteria for Inclusion

In order to keep this project down to a manageable size, I placed the A1 at the centre of a 1-kilometre corridor and then combed the entire 396-mile route, deeming a site worthy of inclusion if its history and design were particularly noteworthy, or if it was simply a striking feature within the roadside landscape. Some names which are synonymous with the A1 failed to make the cut: there is no room for the drab service station and hotel at Scotch Corner, while the Black Cat roundabout is also omitted from the main list due to the eponymous metal feline sculpture resembling something salvaged from a garden centre. A focus on the built environment also means that natural landscape features are not included. Neither are archaeological sites, a specialised subject area which probably merits a treatise in its own right.

While this is not a dedicated history of the A1 road itself, many of the major engineering projects, alignment changes, bypasses and other improvements are described within the text. For a comprehensive account of the route and its innumerable alterations since 1921, may I direct to you to *The Great North Road Then and Now* by Chris Cooper.

Notes on the Text

All of the research for this book was carried out in the period 2011–15, with photographs credited to myself also from this time. I have included the locations for all landmarks so that the reader can find their precise position if they wish: typing the latitude and longitude coordinates into the search box within Google Maps then pressing enter should pinpoint the structure. Please bear in mind that many of the landmarks listed in this book are not open to the public and are situated on private and/or inaccessible land. If visiting a building please do not trespass, and respect the privacy of the owners. Where available I have listed the website applicable to each landmark which often gives details of accessibility.

Landmarks highlighted in italics are described in full elsewhere within the book, with the figure in brackets indicating the number the landmark has been given, 1–80.

Many of the structures featured have been added to the Statutory List of Buildings of Special Architectural or Historic Interest, affording them a measure of protection from alteration or demolition. These 'listed' buildings are categorised according to their significance based on a number of criteria, with Historic England and Scotland using different systems of classification. Within England the following system is used:

Grade I buildings are of exceptional interest; 2.5 per cent (of the total).
Grade II* buildings are particularly important buildings of more than special interest; 5.5 per cent.
Grade II buildings are of special interest, warranting every effort to preserve them; 92 per cent.

Historic Scotland define their categories thus:

Category A: Buildings of national or international importance, either architectural or historic, or fine little-altered examples of some particular period, style or building type. (Approximately 8 per cent of the total).

Category B: Buildings of regional or more than local importance, or major examples of some particular period, style or building type which may have been altered. (Approximately 50 per cent of the total).

Category C: Buildings of local importance, lesser examples of any period, style, or building type, as originally constructed or moderately altered; and simple traditional buildings which group well with other listed buildings. (Approximately 42 per cent of the total).

A Note of Caution

'Don't you go playing near that Great North Road' (Virginia Holland)

Please take care when travelling on the A1!

The A1 is a dangerous road with an unenviable safety record. A mixture of congested and chaotic urban sections, rural single and dual carriageway punctuated by often poorly designed at-grade junctions, it is largely unlit at night and frequented by slow-moving agricultural vehicles. There are blind brows, tight curves, lane gains and lane drops to be negotiated. This book has been designed to be enjoyed from the comfort of the living room rather than glanced at from behind the wheel – treat the A1 with respect and stay safe!

Part One: London

① 200 Aldersgate Street
Postmodernist office block
www.200aldersgate.com
Year of completion: 1992
Architects: Fitzroy Robinson
Location: 51.517947, -0.097425

Anyone wishing to trace the development of the office tower in Britain is well advised to take a visit to London Wall. Each era is represented and competes for attention along this inner-city dual carriageway. Standing across the roundabout from the *Museum of London* [2], 200 Aldersgate Street is representative of the style that has come to be known as postmodernism. Many characteristics of 'PoMo' are present at Fitzroy Robinson's home for top legal firm Clifford Chance: the extensive use of marble as a cladding material, setbacks in place of uniform curtain walling, and references to historical styles that range from subtle to overt.

200 Aldersgate Street is rather more subdued than many of its contemporaries (certainly when compared to Terry Farrell's nearby Alban Gate); its highest floors are not marble clad, which helps to soften the building's impact on its surroundings. The building has recently been refurbished by MoreySmith architects as a multi-tenancy scheme, Clifford Chance having relocated to Docklands. MoreySmith replaced the original cascading glass atrium with a more practical podium, while also completing an internal renovation.

Considerably enlivening the somewhat drab ground level area next to the roundabout are two colourful circular neon artworks by Rob and Nick Carter, which denote the entrances to the building.

② Museum of London
Modernist structure housing gallery and exhibition spaces
www.museumoflondon.org.uk
Year of completion: 1976
Architects: Powell Moya & Partners
Location: 51.517964, -0.096427

With over 400,000 visitors in 2009 (including its site in London's Docklands) the Museum of London is a popular stop on the capital's well-worn tourist trail. Attempting to tell the story of London in an engaging and accessible way was the daunting task set for this amalgamation of the Guildhall and London museums on a new site adjacent to the Barbican development. It was anything but a clean piece of paper design for its architects. Firstly *Ironmongers' Hall* [3], squarely in the middle of the plot, had to be retained. Then the new building needed to be linked up to the system of elevated highwalks that traverse the Square Mile, meaning the entrance to the museum is, confusingly, not at street level. Finally, an office tower entirely separate from the museum but part of the same commission had to be incorporated. Bastion House, last unaltered survivor of the six curtain wall office slabs built along London Wall in the 60s and 70s, perches

awkwardly above on concrete pilotis. Design of the exhibition spaces was left in the hands of Higgins, Ney & Partners.

Critical reaction has generally followed a similar pattern: praise for the variety of exhibits and layout of the gallery spaces, dismay regarding the drab and uninviting exterior elevations. It is hard to argue against the latter conclusion, though there are some pleasing details, notably the brick rotunda containing a garden that forms the roundabout where London Wall meets Aldersgate Street, and the concealed window which allows a view of the remains of the Roman wall outside while visitors examine relics from Roman London within. There are seven separate sections arranged chronologically, each dealing with a different era of the capital's development. Notable architectural exhibits include two salvaged art deco treasures: an iron entrance gate panel from the Firestone factory formerly in Brentford and an ornate lift from Selfridge's department store.

'Union: Horse with Two Disks' by Christopher Le Brun is a bronze sculpture placed on the entrance concourse that provides an irresistible temptation for a few bold visitors, who mount the horse for an impromptu photo opportunity.

3 Ironmongers' Hall
Tudor revival Livery Hall
www.ironmongers.org
Year of completion: 1925
Architect: Sydney Tatchell
Location: 51.518193, -0.096277

Any lost tourists trying to find their way into the *Museum of London* [2] might chance upon this structure and believe they have stumbled across a relic of the sixteenth century. In fact Ironmongers' Hall was built as recently as 1925, as a replacement for a previous building which stood on Fenchurch Street, this being damaged in a 1917 Gotha air raid and subsequently demolished. In its place the Ironmongers' Company (one of the great twelve livery companies in the City of London) chose to start afresh at a site adjacent to the junction of London Wall and Aldersgate Street.

Hidden behind the neo-Tudor elevations is a modern steel frame, with the façade being composed of hand-made Daneshill bricks on the lower levels, then a half-timbered upper storey. Above all this perches a lofty chimney stack. Interiors are opulent, with much wood panelling. A grand banqueting hall and other large rooms provide the space required for the various conferences and events which nowadays help sustain the company.

Since its completion Ironmongers' Hall has twice narrowly escaped destruction, first at the hands of the Luftwaffe in 1940, when much of the surrounding area was destroyed, then in the initial scheme for the Museum of London, which proposed demolition. A far less decorative extension by Fitzroy Robinson & Partners was completed in 1979.

4 Lauderdale Tower, the Barbican
Residential tower in Brutalist style
www.barbican.org.uk
Year of completion: 1974
Architects: Chamberlin, Powell & Bon
Listed status: Grade II
Location: 51.519842, -0.096903

By far the tallest inhabited structure on our list, this is the westernmost of the Barbican's trio of towers, which were once the tallest residential blocks in the UK (that distinction was assumed by One St George Wharf on completion in 2013). Their uncompromising Brutalist aesthetic still dominates the vicinity. Roughly triangular in plan and soaring to forty-four storeys, the Ove Arup-engineered towers are faced in pick-hammered granite and feature daringly cantilevered balconies.

Lauderdale Tower forms part of the huge Barbican complex, which utilised a large tract of land north of St Paul's that was destroyed during the Blitz. It was completed from 1956 to 1981 and features a theatre, art gallery and schools as well as the apartment blocks. Despite a construction period of a quarter of a century, the scheme is remarkably coherent and has survived unaltered. It is the most famous completed project of its architects, who had already designed the adjacent *Golden Lane Estate* [5].

Prospective owners here will need deep pockets as well as a head for heights: flats on the uppermost floors can command prices in excess of £1 million.

5 Great Arthur House, Golden Lane Estate
International Style residential tower block
www.goldenlaneestate.org
Year of completion: 1957
Architects: Chamberlin, Powell & Bon
Listed status: Grade II
Location: 51.522164, -0.095959

Great Arthur House is the most striking element of the Golden Lane Estate, a housing scheme conceived by the City Corporation and executed by the fledgling architectural team of Chamberlin, Powell & Bon, who had submitted separate entries to the competition with the understanding that they would form a practice should one of them win. Geoffry Powell's designs were selected, marking the beginning of a thirty-year programme of construction in this area for which the trio were responsible.

Tasked with designing the sixteen-storey tower, Peter Chamberlin gave each single bedroom flat its own balcony and a modern open-plan layout divided by a sliding partition. Externally, the most distinctive features are the golden-coloured spandrels, which contrast with the blue and red palette found in the rest of the estate, and the remarkable concrete canopy projecting out over the summit of the building. Ostensibly disguising rooftop plant structures, it also gives the estate its signature motif.

For all its architectural merits, Great Arthur House is now showing its age. The single-glazed curtain wall is in urgent need of renewal: close inspection reveals cracked, discoloured panels and leaking aluminium window frames. John Robertson Architects were reported to have been selected to complete this remedial work, which will hopefully preserve the original appearance of the structure.

⑥ The Hat and Feathers public house
Defunct former public house with bowed Italianate frontage
Year of completion: 1860
Architect: William Finch Hill of Finch Hill & Paraire
Listed status: Grade II
Location: 51.523446, -0.098936

With its jaunty neoclassical exterior, this public house provides an ornate backdrop to the drab expanse of tarmac that forms the junction of Goswell and Clerkenwell roads. It is the work of an architectural practice that specialised in designing entertainment venues for the burgeoning Victorian lower-middle classes, notably the Philharmonic Hall in Islington High Street and the Horseshoe Hotel on Tottenham Court Road (both now demolished). Originally built as three separate addresses, The Hat and Feathers now incorporates the one remaining house, which is the simpler adjoining section along Clerkenwell Road. On the pub itself there are some fine decorative elements finished in stucco including elaborate foliage capitals and statues of female figures to the uppermost floor.

The building is currently surrounded on three sides by a surface car park. This valuable plot of land has been coveted by property developers for years, though the pub's location within a conservation area means that the site has thus far escaped any insensitive developments. After standing empty for over a decade it was subject to a comprehensive renovation project costing £1.5 million, with the emphasis being shifted from traditional drinking establishment to

fine dining: a first-floor restaurant was opened which was hoped would recoup some of the investment, however by 2011 The Hat and Feathers closed its doors once again and now faces an uncertain future.

The unusual name is thought to be derived from the hats decorated with ostrich feathers which were worn by Prince Rupert and fellow Cavaliers during the English Civil War.

7 Turnpike House
Concrete-framed residential tower block
Year of completion: 1965
Architect: Carl Ludwig Franck of Franck & Deeks
Location: 51.527400, -0.099905

Dotted around the former Borough of Finsbury are numerous tower blocks, many of which are strikingly similar in appearance. These are the work of C. L. Franck, a German émigré architect who came to Britain in the 1930s, gaining employment with the famous Tecton

group. This practice had been commissioned by the progressive Labour-controlled council to design new amenities for the impoverished borough, including the renowned health centre of 1938. The Second World War interrupted these plans, with widespread destruction of housing stock by German bombers making the requirement for residential development even more urgent. Franck was to continue with this work into the 1960s, first as part of Emberton, Franck & Tardrew, then with Franck & Deeks.

Turnpike House sits at the western end of King Square. This was laid out in the 1820s, though the only surviving element from that period is the church of St Barnabas. From ground level the most noticeable feature of this 60-metre tower is a pair of wide, flat concrete arches which form a gateway to King Square Gardens beyond. Curved entrance arches or rooftop canopies are a recurring motif of Franck's point blocks for Finsbury, with common finishes also applied to the main elevations. At Turnpike House

14

the prefabricated concrete frame is infilled by mosaic, aggregate and fluted concrete panels.

Nearby are two even taller towers from the same design team: Michael Cliffe House soars 74 metres above Skinner Street, while a short distance up Goswell Road Peregrine House tops out at an impressive 80 metres.

8 Former Angel Picture Theatre
Converted tower remnant of former cinema
Year of completion: 1913
Architect: Harry Courtenay Constantine
Listed status: Grade II
Location: 51.532227, -0.106473

The skyline around Islington High Street is pierced by the terracotta cupola of the former Angel Hotel and this Italianate tower, the only surviving element of one of London's earliest cinemas. Its imposing height and classical detailing give an indication of the

opulence that could once be found within the now demolished 1,463-seat auditorium. This featured marble surfaces, stained glass and elaborately decorated capitals beneath a coffered, vaulted roof. Most of the audience would have entered via the main entrance on White Lion Street, with only those with dress circle tickets gaining access from Islington High Street. It was the intention of the developers to screen educational, informative offerings to the local population, though quite how long this policy was maintained is unclear.

Declining cinema audiences spelled the end for the theatre in the early 1970s, with only the tower escaping the wrecking ball. It seemingly being a rule of twenty-first-century London that no-one should ever be further than 50 metres from a coffee shop, this now houses a branch of Starbucks on its ground floor. An intrepid graffiti artist has somehow managed to reach the second stage of the tower and add their tag in blue and yellow aerosol spray, this having escaped being buffed at the time of writing.

9 Business Design Centre

Remodelled exhibition and conference centre
www.businessdesigncentre.co.uk
Year of completion: 1862
Architect: Frederick Peck / Ironwork: Andrew Handyside & Co.
Listed status: Grade II
Location: 51.535645, -0.106094

Now that the streets of Islington are overrun with Porsche Boxsters and Mini Coopers, it is difficult to imagine an era when these same roads would often have been crowded with livestock on their way to Smithfield Market. Members of the Smithfield Club realised the area would be ideally suited for a grand hall in which cattle and the latest agricultural machinery could be displayed. By using wrought iron, a huge obstruction-free covered space was created, albeit one that would soon be eclipsed by the even larger spans built at St Pancras railway station and the National Agricultural Hall (Olympia).

Besides the Smithfield Show, the huge new building proved to be adaptable enough to host two other famous annual events: the Crufts dog show and the Royal Military Tournament, before being requisitioned by the General Post Office as a parcel depot during the Second World War. It was to remain in use by the GPO until 1970. By the time it made an appearance in an episode of *The Professionals* in 1980, 'The Aggie', as it was known, was a derelict hulk. It was only rescued by the initiative of businessman Sam Morris, who saw its potential as a modern trade and exhibition centre. Architects Renton Howard Wood Levin were commissioned to undertake a comprehensive redevelopment of the site.

On reopening in 1986, the renamed Business Design Centre had been transformed. Several adjoining buildings had been demolished, the impressive old entrance on Liverpool Road replaced by a new one to Upper Street (unmistakeably 80s with its series of barrel vaults) and a mezzanine floor added to the main hall. Restoration of the roof resulted in the loss of the original glazing, replaced by polycarbonate sheeting.

10 St Mary, Islington
Parish church with eighteenth-century tower and steeple
www.stmaryislington.org
Year of completion: 1754 (tower) 1956 (nave)
Architects: Lancelot Dowbiggin (eighteenth-century building), John
Seely & Paul Paget (1956 reconstruction)
Listed status: Grade II
Location: 51.538355, -0.101784

A tactical decision made by Hitler and the Luftwaffe high command
was to have fateful consequences for St Mary's. Switching the
Luftwaffe's attention from daylight raids on RAF airfields to
night-time sorties against the capital, with a consequent loss of
bombing accuracy, resulted in the nave being damaged beyond repair
by a high explosive bomb on the night of 9 September 1940. It would
be more than a decade before services could resume at the site.

Three distinct eras of construction comprise St Mary's as seen
today. Oldest and most prominent are the tower and steeple,

all that remains of the Georgian church
following the wartime bomb damage. Rich
in classical detail, this features a tapering,
tiered spirelet pierced by oval openings
which is supported by an open circular
stage. Below this the brick tower with large
clock faces to three sides. At the west end
facing Upper Street is the entrance portico,
added in 1903: paired Ionic columns beneath
a segmental pediment. Finally the re-built
nave, which replicates the brick with stone
quoins seen on the steeple but is stripped
of ornamentation, both inside and out. Tall
oblong windows flood the interior with
light, illuminating a depiction of Christ by
Brian Thomas.

A desire to foster greater links with the
local community led to the refurbishment of
the crypt in 2009, which can now be hired out
for meetings and events.

⑪ Former Royal Mail public house
Converted Victorian-era public house
www.hoxleyandporter.co.uk
Year of completion: 1879
Builders: Francis Bradley & Son
Listed status: Grade II
Location: 51.540357, -0.102814

An intriguing little building that is squeezed into a narrow plot on busy Upper Street. There has been an inn on this site since the seventeenth century (a stuccoed shield on the second floor bears the date AD 1631) with the present structure dating from 1879. A painting of 1840 depicts its predecessor, which featured a two-bay-wide ground floor entrance beneath a segmentally pedimented cornice. This was replaced by the present building with its noteworthy projecting bay window and flanking doorways, all finished in decorative timber panelling. Above this are two floors completed in brick and stucco with dividing pilasters, a dentil cornice and a triangular central pediment. The name presumably relates to the building's location on the old mail coach route heading north.

In recent years the Royal Mail name has been dropped as several different owners have attempted to attract their share of the competitive local food and drink trade. At one time known as the Jorene Celeste, then the Grand Union, as of 2014 the property is now home to the Hoxley and Porter bar/restaurant. This offers steak frites at £12 among other gastropub dishes in a transformed interior, which is distinguished by patterned wallpapers that can politely be described as 'vibrant'.

⑫ K2 telephone kiosk
Cast iron telephone box in neo-classical style
Year of design: 1924
Designer: Sir Giles Gilbert Scott
Listed status: Grade II
Location: 51.544813, -0.103064

This example of a K2 kiosk is located adjacent to the entrance of the *Union Chapel* [13] on Upper Street. It is one of more than 200 such models to have been given protected status following a successful campaign by the Thirties Society (now the Twentieth Century Society): they were to have been swept away in a wave of modernisation following the privatisation of British Telecom in 1984.

Its designer was more accustomed to building on a far grander scale, being responsible for Battersea Power Station and Liverpool Anglican Cathedral during a prolific and diverse career. His design for the K2 resulted from a competition held by the Royal Fine Art Commission in which Scott's concept was declared the winner. Its form is said to have been influenced by the work of Sir John Soane, with the ceiling of the breakfast room at Soane's residence (now a museum) and the self-designed family tomb both being plausibly cited.

Constructed of heavyweight cast iron, the K2 proved too expensive to be adopted nationwide; nearly all were erected within London. Scott had suggested silver for the exterior colour but was overruled by the GPO, though he was invited to design a lighter, smaller version which could be produced more economically. This became the K6. It's only when the two are placed together that the difference in size becomes obvious, with the K2 towering over its later sibling.

⑬ Union Chapel
Gothic nonconformist chapel set within Georgian terrace
www.unionchapel.org.uk
Year of completion: 1877
Architect: James Cubitt
Listed status: Grade I
Location: 51.544759, -0.102524

One of the finest religious buildings in the capital. Like its near contemporary the Midland Grand Hotel at St Pancras station, it was once seriously threatened with demolition, a testament to how architectural styles can fall in and out of fashion over the generations. Now it is once again a valued community asset, with the building doubling as a music venue. In recent years, acts as diverse as Adele, Goldfrapp and Nick Lowe have performed under the blind lancets of the magnificent octagonal wooden ceiling.

Two men were instrumental in shaping the appearance of Union Chapel: Dr Henry Allon, pastor from 1844 until his death in 1892, and architect James Cubitt, author of *Church Designs for Congregations* and noted designer of nonconformist churches. In his book Cubitt highlighted the inadequacies of the traditional 'Gothic' church design (i.e. an oblong nave with aisles), claiming that a large proportion of a congregation were unable to either fully see or hear the celebrant. He found a kindred spirit in Dr Allon, who demanded clear sight lines between pulpit and pews with excellent acoustics when it became necessary to rebuild the original neoclassical Congregationalist chapel on Compton Terrace. Islington was expanding rapidly during this period and more seating space was required. Among the worshippers at this time was a young H. H. Asquith, later Liberal prime minister.

Outstanding examples of lavish Victorian craftsmanship are apparent throughout. An imposing 170-foot tower completed in red brick with stone dressings is the focal point of the exterior. This was completed twelve years after the rest of the building. Inside the central space there is capacity for 1,600 worshippers on wooden pews of Colombian pine. These face a stone pulpit with marble detailing carved by Thomas Earp. Hidden by a screen behind this lies a restored Henry Willis organ which retains its original hydraulic blowing system, believed to be the only surviving example in the country. Both this and the stained glass rose window above (which features angels playing musical instruments) illustrate the importance Dr Allon placed on music as an integral part of any religious service.

14 St Mary Magdalene, Holloway
Neoclassical parish church within public garden
www.hopechurchislington.org
Year of completion: 1814
Architect: William Wickings / Builder: Joseph Griffiths
Listed status: Grade II*
Location: 51.548418, -0.107843

On a warm summer's day, St Mary's church gardens offer a welcome relief from the hectic activity along Holloway Road. The expansive former churchyard is shaded by mature Lime and London Plane trees and is a popular spot for the local community to relax in. Within it, the rather austere neoclassical Regency church was originally built as a chapel of ease to *St Mary, Islington* [10]. It became a parish church in its own right in 1894. Constructed from yellow stock brick, there is a notable absence of ornamentation to the exterior (due no doubt to financial constraints) though the squat east tower is distinguished by stone Ionic pilasters above which there is a balustraded parapet, with paired urns to each corner.

The interior has been much altered over the 200 years of the church's existence. The rectangular gallery is a late Victorian modification from the original horseshoe shape. On the ground floor the original pews and choir stalls have been removed while the spaces beneath the galleries have been filled in and are used as meeting rooms. Hope Church Islington's aim is to function as a hub of the local community; to this end there are a range of social events, courses and groups to attend as well as a variety of services.

⑮ Islington Central Library
Early example of 'open-access' public library
Year of completion: 1907
Architect: Henry T. Hare
Listed status: Grade II
Location: 51.549137, -0.107400

Following exponential population growth through much of the nineteenth century, it was somewhat belatedly decided to provide the inhabitants of Islington with a network of public libraries. More than half a century after it was first introduced, the Public Libraries Act was finally adopted, meaning that the borough council could levy funds via ratepayers. This in itself would not have been enough to construct a building as lavish as that which was opened in October 1907 and stands on the corner of Fieldway Crescent. A flamboyant Edwardian baroque structure faced mainly in Portland stone, it owes its existence to the philanthropic actions of the industrialist Andrew Carnegie. Having sold his American steel company for a vast sum, Carnegie then dedicated his life to providing facilities for learning for the underprivileged. The people of Islington were among the beneficiaries of Carnegie's largesse: £40,000 was donated in total, of which half that amount was spent on the central library.

With its statues of Spenser and Bacon gazing down on Holloway Road, this library stands as the epitome of the Victorian/Edwardian ideal of providing imposing civic buildings that would benefit an 'improving' (and potentially volatile) urban working class population, while reflecting well on those that had provided it. There was also a notable innovation in the new Islington libraries: instead of having to request reading material from a storage repository accessible only by staff members, these new 'open-access' designs allowed visitors to browse books directly from the shelves before checking them out. It is a system we take for granted today, but which would only become widespread in the inter-war years.

A requirement for more floor space led to the construction of an extension in a plain modern style, completed in 1976. This is now where the lending library is sited, with the original building's only public function being as the location of a First Steps Learning Centre.

⑯ London Metropolitan University Holloway Road Hub
Higher Education campus in urban setting
www.londonmet.ac.uk
Year of completion: 1896 (Northern Polytechnic Institution Building), 1966 (Tower Building), 1994 (Learning Centre), 2000 (Technology Tower), 2004 (Graduate Centre)
Architects: Charles Bell (Northern Polytechnic Institution Building), London County Council Architects – Schools Division (Tower Building), Geoffrey Kidd Associates (Learning Centre), Rick Mather Architects (Technology Tower), Studio Libeskind (Graduate Centre)
Location: 51.552136, -0.111108

Quite a convoluted history of mergers and rebrandings for this troubled and perennially underperforming higher education institution. First opening its doors in Holloway as the Northern Polytechnic Institution back in October 1896, it was the subject of a merger with the North Western Polytechnic in 1971, becoming the Polytechnic of North London. Conferment of university status in 1992 then led to the adoption of the title University of North London. Finally (for now at least) there was a further unification, this time with London Guildhall University, which led to the creation of London Metropolitan University in 2002.

As might be expected from a campus that has been developed in an ad hoc manner for over a century, architectural styles are many and varied. Neoclassicism, brutalism, postmodernism and

deconstructivism are all represented. At some point the façade of Charles Bell's original institution building has been denuded of its ornate pediments, keystones, dentilled cornice and balustraded parapet, though the prominent clock tower has survived. To the north of this there is the overpowering presence of the Tower Building, thirteen storeys rising to 46 metres and finished in pre-cast concrete cladding panels. There are distinct similarities in appearance with Guy's Hospital Tower at London Bridge, albeit on a smaller scale. It is now partially obscured by Rick Mather Architects' Technology Tower, its curved frontage already looking grey and weathered.

No doubt an awareness of LMU's relative lack of prestige along with its poor performance in university league tables contributed to the decision to hire a 'starchitect' capable of producing a design that would act as a distinctive landmark and as an emblem of a more self-confident university. Daniel Libeskind's Graduate Centre gives the appearance of having crash-landed in Holloway Road and then sunk halfway into the ground. It is composed of three intersecting volumes clad in stainless steel panels with doorways and fenestration leaning at crazy angles both inside and out, as if the architect had been inspired by watching the German Expressionist classic *The Cabinet of Dr Caligari*. Perhaps a cinematic inspiration is appropriate here as the site was once occupied by the Holloway Grand Pictures cinema. Winner of a RIBA award in 2004, it is nevertheless debatable how successful an addition to the cluttered urban streetscape of Holloway this building is: it would arguably work better with more space around it, perhaps as the centrepiece of a city plaza.

17 Holloway Road Station
London Underground station clad in faience
Year of completion: 1906
Architect: Leslie W. Green
Listed status: Grade II
Location: 51.552773, -0.112801

Any pedestrian that glances up on their way to catch a train may be puzzled by a cryptic set of initials occupying the northernmost bay of this building: GNP&BRY. This is the Great Northern, Piccadilly & Brompton Railway Company, which was one of the lines operated by the Underground Electric Railways Company of London (UERL). Development of the integrated system of underground routes we are familiar with today was initially the domain of competitive private enterprise. By the time of his death in 1905 American entrepreneur Charles Tyson Yerkes had acquired several undercapitalised railways and merged them under the UERL umbrella. One of these lines was the GNP&BR, which today forms part of the Piccadilly line.

Holloway Road is typical of the many station buildings designed by Leslie Green for Yerkes. Though by necessity variations on a theme tailored to individual sites, they nearly all featured façades clad in 'ox-blood' faience (glazed terracotta) blocks, semi-circular windows above the entry and exit portals and tiled interiors in contrasting colours. These finishes were no doubt chosen for their hard-wearing qualities and have stood up well to a century of use. At Holloway Road much of the interior detailing has survived, including the charming art nouveau ticket booths and the original platform signage.

⑱ Odeon Holloway
Former Gaumont cinema with opulent original foyer
www.odeon.co.uk/cinemas/holloway/97/
Year of completion: 1938
Architects: C. Howard Crane (auditorium reconstruction by T. P. Bennett & Son, 1958)
Listed status: Grade II
Location: 51.558509, -0.121021

Occupying a prominent corner site at the junction of Holloway and Tufnell Park roads, this Odeon began life as one of the super-cinemas constructed for the Gaumont-British Picture Corporation. With a capacity of over 3,000, all seated in one huge tiered auditorium, this was a true picture palace designed to entice the audiences which were then flocking to the movies in their millions. Attendances in Britain would peak shortly after the war, by which time the Holloway Gaumont was a tangled ruin, its lavish interior wrecked by a V1 flying bomb. It was not until 1958 that the refurbished cinema was reopened, having been reduced in capacity and, foyer excluded, the interior redesigned in a pared-back modern style. Since then further alterations have resulted in the creation of eight downsized screen rooms.

Twenty-first-century audiences used to the joyless surroundings of the contemporary multiplex would probably find it difficult to compare their modern cinematic experience with those who bought a ticket for the opening night screening of *The Hurricane* starring Dorothy Lamour, back in 1938. Approached from Holloway Road, the first impression would have been of the imposing size of the faience-faced corner tower, adorned with decorative friezes and arabesques. Once inside the foyer patrons could then absorb an interior where every surface was richly decorated. At one end was a sweeping staircase, flanked by fluted Corinthian pilasters lining the walls, and the whole scene lit by chandeliers. Inside the main auditorium it was no less luxurious and continued the ornate Renaissance theme, with the screen itself framed by a gilded proscenium.

Many of the decorative elements that survived the aerial attack have since been hidden behind partitions as the building has been adapted. However the foyer remains in a relatively original condition, allowing modern cinephiles a glimpse of the glamour afforded to a previous generation of cinema-goers.

⑲ St John the Evangelist

Gothic revival Commissioners' church in urban setting
www.stjohnsevangelist.co.uk
Year of completion: 1828
Architect: Sir Charles Barry
Listed status: Grade II*
Location: 51.563841, -0.130274

One of many ecclesiastical buildings from this period that owes its existence to a series of Acts of Parliament, the first of 1818. These created a body of Church Building Commissioners who could allocate grants towards the construction of new churches from an initial fund of £1,000,000 (this first Act became known as the 'Million Act'). Both Church and state were anxious to provide the rapidly expanding urban populations with places of worship to head off the perceived threats of Nonconformism and militancy within the working classes. There was a distinct emphasis on speed and economy of construction, with many of the new structures featuring a simplified 'austerity Gothic' made necessary by limited budgets.

When completed, St John would still have been in largely open country north of London, near to the village of Highgate. Within a few decades any remaining undeveloped land would quickly be swallowed by dense residential and industrial agglomerations. As church attendances dwindled away in the twentieth century St John was amalgamated with nearby St Peter's, which has been converted to residential flats.

Barry is most famous for designing (with A. W. N. Pugin) the Houses of Parliament. This is one of his lesser works, and has generally been derided by architectural historians. It has a slim west tower topped with pinnacles which are replicated on the buttresses at both ends of the building. The interior features some rare box pews in addition to a wooden gallery. Both the stained glass in the east window and the chancel fittings are believed to be later additions.

20 Archway Tower
International Style office tower above Underground station
Year of completion: 1963
Architects: Oscar Garry & Partners
Location: 51.565539, -0.134988

An attempt to bring the sleek style of pioneering New York skyscrapers such as the UN Secretariat building to north London, but executed without any of the panache on a restricted budget, which precluded the use of high-quality cladding materials. It is a formulaic slab-on-podium design consisting of three slender towers clustered together and braced by blank end walls. These are clad with pre-cast aggregate panels, while the curtain walls which comprise the main elevations are a rhythm of blue-grey tinted glass and matching spandrels.

Considering Archway Tower is now more than fifty years old, its fabric has aged reasonably well; its problem has always been how it relates to the street level area around its base. A dismal environment for the pedestrian, it is blighted by heavy traffic as well as by strong gusts of wind funnelled down to ground level by the tower. The area is crying out for investment, which may have finally arrived if plans for the tower and the road layout come to fruition. Property developers Essential Living have acquired the vacant building and plan to convert it to residential flats. This will involve a complete renewal of the façade and is scheduled for completion in 2016. By this time work may also have started on converting the dangerous 'gyratory' system for the benefit of pedestrians, cyclists and those using public transport.

'How to Make Archway Tower Disappear' was the title of a 2012 temporary installation by Ruth Ewan on Holloway Road. This used a viewer incorporating specially designed computer software to allow members of the public to see a live image of Archway from street level, only with the 59-metre tower magically removed from the field of view. Many local residents would no doubt be delighted if this temporary 'disappearance' could be made permanent.

㉑ Highgate Archway
Wrought iron bridge carrying Hornsey Lane over Archway Road
www.hornseylanebridge.net
Year of completion: 1900
Architect: Sir Alexander Binnie
Listed status: Grade II
Location: 51.570618, -0.138362

This bridge provides a wonderful vantage point to look back towards central London, with most of the capital's tall buildings being discernible including *200 Aldersgate Street* [1], *Lauderdale Tower* [4] and *Turnpike House* [7]. Highgate Archway was built as a replacement for an aqueduct-style structure designed by John Nash, which proved impractically narrow for the increased traffic numbers utilising this new route out of London. Previously vehicles would have had to traverse the steep hill through Highgate. Rising 60 feet above the dual carriageway, the road deck is supported by a segmental arch composed of seven girders, which are anchored to masonry abutments. This is not a purely utilitarian structure: the spandrels and cornice feature rich classical decoration while the lamp standards that line the parapet are of a similar design of those that can be seen on the Thames Embankment. They are by the architect George Vulliamy.

Unfortunately Highgate Archway has gained notoriety in recent years as a location for suicide attempts. In 2010 alone three men

jumped to their deaths within three weeks, with further fatalities since. This has led to pleas from the local community for greater steps to be taken to prevent these tragedies. Haringey Council has so far refused to sanction the deployment of a safety net below the bridge, which campaigners believe would help to ensure that no more lives are lost on 'suicide bridge', as it has become known.

㉒ St Augustine of Canterbury
Remodelled Victorian Gothic parish church
www.saintaugustine.org.uk
Year of completion: 1896
Architects: John Dando Sedding / J. Harold Gibbons (west front)
Listed status: Grade II
Location: 51.572117, -0.139490

Three phases of development characterise this very distinctive church, built to serve the expanding community along Archway Road: Sedding's original eight-bay nave, Gibbons' flamboyant west front added in 1916, and the same architect's modifications to the east end dating from 1925. Sedding was a noted luminary of the Arts and Crafts movement who designed both new ecclesiastical buildings and sensitively restored existing ones. His Grade II* listed Holy Redeemer church in nearby Finsbury is noteworthy for its Italianate exterior (Sedding usually favoured derivations of the Gothic style) and innovative steel frame. Many of Sedding's buildings were completed after his death by his assistant Henry Wilson.

This original structure was soon partly obscured by the construction of a new west front, finished in yellow stock brick with stone dressings. An unusual louvred pagoda roof sits above a heavily buttressed tower, this containing a large west window distinguished by flowing organic tracery. Gibbons made further alterations following a major fire in 1924, this time to the east end where an organ gallery replaced three of the original pointed clerestory arches.

㉓ Belvedere Court
Block of modernist residential flats
Year of completion: 1938
Architect: Ernst Freud
Listed status: Grade II
Location: 51.584622, -0.169900

Youngest son of Sigmund, father to Lucian and Sir Clement, Ernst Freud was compelled to seek refuge from the advance of National Socialism in Germany, arriving in London from Berlin in 1933. This building is a contemporary and north London near-neighbour of Erno Goldfinger's row of houses on Willow Road, Wells Coates' Lawn Road Flats and Berthold Lubetkin's Highpoint 1 and 2, all icons of the Modern Movement in architecture. Belvedere Court has never gained such an exalted reputation among architectural historians. Perhaps this is because Freud omitted some of the more radical visual features of modernism for a more conservative British audience: there are bare brick walls rather than white render, while instead of a flat roof there is a slightly incongruous mansard attic. Perhaps the most pleasing view can be gained from the grass bank immediately to the west of the building, from where the streamlined pavilions align themselves in a stepped back formation.

㉔ Finchley Synagogue
Large reinforced concrete suburban synagogue
www.kinloss.org.uk
Year of completion: 1967
Architects: Dowton & Hurst
Location: 51.590328, -0.201626

As the Jewish community migrated from its traditional stronghold in the East End to the suburbs, so the London United Synagogue accelerated its building programme to cater for this rapidly redistributing population, with the largest of these new shuls able to comfortably accommodate over 1,000 worshippers. Finchley was admitted as a district synagogue in 1935, establishing itself on Kinloss Gardens in a modest brick building with pitched roof designed by Cecil J. Eprile, with room for 500 people. By the beginning of the 1960s it was evident that rising membership would necessitate further expansion. At a cost of £400,000, a new reinforced concrete structure rose on the same plot with its distinctive trademark 'radiator grille' façade facing the busy A1/A406 multiplex. Its grand opening in 1967 was attended by the local MP, a rising star in the Conservative Party by the name of Margaret Thatcher.

Inside, Finchley largely conforms to the pattern of the other 'cathedral synagogues' from this era, with a galleried interior lit by clerestories and a capacity of 1,320. The Jerusalem Window features stained glass by R. L. Rothschild. Despite these extensive and often lavish construction projects, only one post-war London synagogue has thus far been listed: Marble Arch by T. P. Bennett & Son (1961).

Part Two: Southern England and the Midlands

㉕ The Galleria
Outlet shopping centre built above road tunnel
Year of completion: 1991
Architects: Michael Aukett and Ken Simms
Location: 51.760851, -0.240792

A slice of American mall culture transplanted to the dormitory town of Hatfield in deepest Hertfordshire. It occupies a vast site straddling the southern end of the Hatfield Tunnel, completed in 1986 to relieve a traffic choke point where the A1 had not been upgraded to motorway status. A 1,150-metre-long cut-and-cover tunnel was employed – this solution having the advantages of limiting noise pollution while creating a tract of land that could then be developed. What eventually emerged was an ambitious £200 million scheme that linked shopping and entertainment areas via a covered walkway. Its most striking visual element is the central concourse, which is covered by a cable-stayed barrel roof. This vault is supported by inclined masts which ensure that the tunnel roof below does not become a weight-bearing element.

Besides this engineering feat, The Galleria is at best forgettable visually. Shop units, car parking and the nine-screen Odeon 'digiplex' cinema are housed in anonymous grey sheds. These will hopefully be partially screened in future as the trees planted around the site perimeter reach maturity. Along with these aesthetic considerations, the complex is often cited as a prime factor in the decline of nearby Hatfield town centre as a shopping destination.

26 KFC fast food outlet and adjacent gatehouse

Moderne former personnel office and gatehouse to de Havilland works

Year of completion: 1934
Architect: Geoffrey Monro
Listed status: Grade II
Location: 51.764948, -0.238137

Two of a quartet of buildings that line the route of a previous A1 alignment (now the A1001 Comet Way) that formed the entrance to the huge de Havilland works and adjoining airfield (see also *Hatfield Police Station* [27]). They now lie out of sight to travellers on the A1, who nowadays traverse Hatfield via a 1.2-km tunnel underneath *the Galleria* [25]. This is a pity, as it denies a wider audience a glimpse of these well preserved examples of 1930s streamlined moderne. They are both of concrete construction with flat roofs, metal glazing bars to the curved windows and feature extensive use of glazed bricks and tiles in an evocative mint green.

Following the end of manufacturing activity at British Aerospace in 1993, these buildings were left to decay. However both have recently been restored and adapted: the former gatehouse now houses a hairdresser's, while the personnel block has been transformed into an outlet for a well-known purveyor of fried chicken. While this conversion employs notably more subtle signage than is usual, it is still difficult to imagine Sir Geoffrey de Havilland strolling to the counter and ordering a Zinger Tower Burger. Such is progress.

㉗ Hatfield Police Station
Former de Havilland offices and staff mess linked by purpose-built criminal justice facility
Year of completion: 1934 (administration block and staff mess) 2008 (police station)
Architects: Geoffrey Monro (1934 buildings) Vincent & Gorbing (2008 restoration and additions)
Listed status: Grade II
Location: 51.765532, -0.238054

Having outgrown their original home at Stag Lane in north London, de Havilland acquired land to the west of Hatfield which provided ample space for manufacturing and flight testing facilities. Most of the structures were sheds and hangars of fairly utilitarian appearance, the exceptions being the buildings constructed either side of the main entrance. Fittingly for a concern engaged in cutting-edge design, they were of clean, modern appearance and were fabricated from reinforced concrete. With its contrasting colour scheme, strong horizontal emphasis and symmetrical plan, the administration block would doubtless have impressed passing motorists on the A1. A neighbouring staff canteen was also completed to cater for the large workforce.

In their 1940s heyday the design offices at de Havilland were responsible for some of the most radical and efficient aircraft ever manufactured. These included the world's first jet airliner, the D.H.106 Comet, an early turbojet fighter (the D.H.100 Vampire) and arguably best of all, the formidable D.H.103 Hornet. This was the successor to

35

the much better known D.H.98 Mosquito, the development of which did not actually take place at Hatfield at all: the design team had decamped to the more peaceful surroundings of Salisbury Hall a few miles away which, once war was declared, was at less risk of air attack than the main factory complex.

De Havilland was eventually absorbed into the Hawker Siddeley group, which itself would become a part of nationalised British Aerospace. Following the end of manufacturing at Hatfield the land was sold off and the site mostly cleared for redevelopment, leaving only a few reminders of its former use. Taking its place is a depressing combination of soulless business park, timid neo-Georgian housing estate and prison-issue university halls of residence. The former staff mess and offices now form part of a police station and criminal justice facility, sandwiching a new reception building featuring inclined glass curtain walls and an oversailing roof canopy.

28 Tudor Oaks Lodge
Timber-framed converted house forming part of hotel
www.tudoroakslodge.com
Listed status: Grade II
Location: 52.030071, -0.215393

This stretch of the A1 was upgraded to dual carriageway in the early 1960s, meaning that northbound traffic now flows perilously closely to this converted house, the ground floor of which is given over to a bar/restaurant. Take a window seat for a slightly unnerving view of articulated lorries thundering past a few feet from your left elbow. Incorporated into a hotel development c. 1980, this is a genuine relic of the Tudor age, though the hotel website's claims of fifteenth-century provenance are probably spurious. Its layout is of the hall with gabled cross-wings type, though there are several modern additions to front and rear. Prominent is an offset chimney stack with octagonal shafts, while there is a mixture of infill to the diagonally braced timber frame – plaster to the front and sides with stretcher bond brick to the rear upper storey. Less satisfying is the presence of a steel extractor flue and several uPVC windows.

Exactly who this house was built for and when is a matter of conjecture; it may have been for a local yeoman. Timber-framed construction was still widespread well into the seventeenth century so it may date from that period. In later years the building became a coaching inn, being ideally located on the Great North Road. It is recorded as the

New Inn on a map of 1808 and has also been known as The Greyhound. Confusingly, there was another pub that was also known as the New Inn close by to the south, this having stood where the northbound BP filling station is located today. By the 1900s it was owned by the Green family before being converted to a farmhouse (with all the half-timbering rendered over) later in the twentieth century. Nowadays it is part of an enlarged business which includes a fourteen-room hotel based around a courtyard and a function room, both of which are finished in an appropriate Tudor revival style.

㉙ Topler's Hill water tower
Reinforced concrete water tower
Year of completion: 1934
Design: Binnie, Deacon & Gourley / Contractor: Peter Lind & Company
Location: 52.047110, -0.229208

This rather forbidding looking ferroconcrete structure appears robust enough to last for hundreds of years to come. It was built for the Biggleswade Water Board by the same firm that manufactured the concrete caissons that formed the temporary Mulberry harbours required for the D-Day landings. Peter Lind & Company were also responsible for the building of the Post Office Tower and the Shredded Wheat grain silos at Welwyn Garden City, which pioneered the slip form construction technique in Britain. Overall cost was £14,500, with the unusual hemispherical metal tank having a capacity of

300,000 gallons. The original steel cap was replaced by an aluminium version in 1980.

Designed to provide water for the villages of Stotfold and Arlesey, the tower stores water pumped from the nearby New Spring pumping station and helps to maintain a constant pressure in the local supply system. There is also a covered reservoir on the site. While there are some notable examples of water towers employing decorative architectural features ('Jumbo' in Colchester and the House in the Clouds at Thorpeness being two of the most famous), here it is a case of form following function, with the upper part of the structure supported by external buttresses.

30 Tempsford Bridge
Arched bridge carrying the A1 over the River Great Ouse
Year of completion: 1820
Architect: James Savage
Listed status: Grade II & Scheduled ancient monument
Location: 52.176976, -0.302242

This venerable structure has been conveying traffic over the River Great Ouse for the best part of two centuries. A replacement for an earlier wooden bridge, this is a simple three-arch design flanked by flood bridges to either side. It was constructed from local sandstone with the cutwaters and arches finished in more durable Bramley Fall stone. Just 10 metres wide, this crossing was soon found to be

inadequate once the age of mass car ownership dawned. When this section of the A1 was 'dualled' in the early 1960s, it was supplemented by a second bridge (a functional concrete design) which was used by southbound traffic, with vehicles heading north utilising the existing crossing. It was intended that the new bridge would eventually be doubled in width, allowing the old section of road to be de-trunked, a plan which has never been implemented.

Despite damage to the parapet caused by wayward goods vehicles and extensive spalling of the sandstone facings, Tempsford Bridge will probably have to remain in service for many years to come, symbolic of the piecemeal development of the A1.

31 St Lawrence, Diddington
Medieval parish church with brick tower
Listed status: Grade II*
Location: 52.278863, -0.256485

Situated along a narrow track, all is tranquil in the vicinity of St Lawrence save for the background hum of traffic from the A1. As you approach the church a boundary wall containing some wrought iron gates is visible on the right. This is the only reminder, other than the mature parkland setting, of the former Diddington Hall. This was demolished in the 1960s. A church is mentioned at Diddington in the Domesday Book, however the current structure is largely early sixteenth-century, built around a thirteenth-century core of nave and chancel, these being finished mostly in an uneven pebblestone. The Tudor-era additions are an early example of the use of brick in a religious building. These would have been hand-cut and are arranged (mostly) in English bond, i.e. alternating courses of headers and stretchers. Also at this time the chancel was truncated, with its east wall completed in brick.

Moving inside reveals more fragments of the church as originally constructed, in particular the round columns and chamfered arches of the north aisle and the octagonal font. There are also some fine sixteenth-century pews with mythical animal carvings.

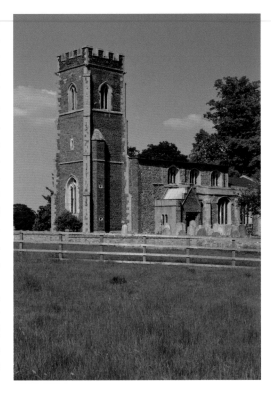

32 The Great Tower and St Mary's parish church, Buckden
Brick tower remnant of bishop's palace / perpendicular parish church
www.stmarysbuckden.org.uk, www.fobt.org.uk
Year of completion: Tower *c.* 1495, Church *c.* 1485
Listed status: both Grade I
Location: 52.294156, -0.252831

Two prominent landmarks that stand cheek by jowl within the village of Buckden on the old Great North Road. The external appearance of both is little changed from the late fifteenth century. Their construction was made possible by the power and wealth of the Diocese of Lincoln, which at one point extended from the Humber to the Thames. Being roughly equidistant between London and Lincoln, this was a convenient location to construct a moated palace that could accommodate the bishop and his entourage as they travelled to and from the capital. It also housed Catherine of Aragon during 1533/34 following the annulment of her marriage to Henry VIII. She was quartered in the Great Chamber, which was

demolished along with most of the other medieval buildings in the nineteenth century. Reduced to little more than a derelict shell, the Great Tower was eventually restored in 1958 after the complex had been acquired by the Claretian Missionaries. With the floors and roof renewed, the tower is now used as a dormitory for visiting students.

St Mary's offers a pleasing contrast with its pale limestone, though both buildings feature similar ornamental castellation around the parapet. Its external appearance is almost uniformly perpendicular, reflecting the rebuilding carried out c. 1435–40 under the auspices of Bishop Gray, then Bishop Alnwick of Lincoln. There is some evidence of the earlier Norman church that formed the basis for this. Entering the nave via the south porch reveals the stone columns and pointed arch of the original doorway. Also dating from the thirteenth century are the piscina (a stone basin) and the stepped seats of the sedilia. These are located in the chancel.

Describing themselves as 'The A1 Church', congregations at St Mary's regularly pray for the safety of motorists and remember those who have lost their lives while travelling on this road.

33 Milestones
Surviving roadside distance markers
www.milestonesociety.co.uk
Listed status: mostly Grade II, Alconbury Grade II*
Location: 52.388361, -0.258257

Of all the listed structures that line the route of the A1, milestones are among the most numerous and the most easily overlooked. They survive mostly where the original alignment of the Great North Road is followed, many of these being relics from the turnpike era. As unpopular then as any proposals for toll roads are now, Acts of Parliament had allowed local trusts to levy a charge on users of their section of highway. Revenues were then to be assigned to road maintenance. By 1766 these trusts were compelled to provide milestones along the entirety of a turnpike road, these being considered essential for the accurate timing of mail coach services. Initially built from stone, they were supplemented by examples cast from iron as the nineteenth century progressed.

Only a fraction of the number of milestones erected remain extant today. Thousands were removed at the beginning of the Second World War as the threat of a German invasion loomed, in an attempt to confuse enemy troops as to their exact whereabouts. Many more disappeared underneath widened roads. Others have been damaged by the elements or simply neglected. However, a fine example has survived in a largely original state at Alconbury Hill: it once marked the intersection of the Old North and Great North roads in Cambridgeshire. Inscriptions on the stone faces display the distance to London depending on which fork in the road the traveller was intending to take. Nearby villages such as Buckden and Stilton are also mentioned. Nowadays it is sandwiched between the six lanes of the modern A1(M) and a huge distribution centre.

34 Sibson Inn hotel
Former coaching inn and farmhouse
www.sibsoninn.com
Year of completion: 1764
Stonemason: Thomas Thompson
Listed status: Grade II
Location: 52.564121, -0.383769

Two and a half centuries have passed since the first of this group of buildings was completed, yet in their attractive honey-coloured oolitic limestone, they remain an inviting sight to weary travellers heading

north on the A1. Originally known as The Wheatsheaf, it was built for the Duke of Bedford as part of his Wansford estate. By the late eighteenth century the site was functioning as a farm; an Ordnance Survey map of 1887 records it as Roadside Farm, with the layout of the buildings much as they remain today. There have been various extensions and additions over the years, with the oldest element being the southernmost sections of the inn, which has two bay windows flanking the doorway. Opposite this at the yard entrance is a barn from the same period.

There is one interesting earlier survival, a stone mounting block and milestone dated 1703. Found outside inns during the coaching era, these blocks allowed for an easier and more dignified mounting of a horse. On this can be found the initials 'E. B.' for Edmund Boulter, a wealthy businessman who provided the funds for a series of these devices to be installed along the Great North Road. The date is believed to be incorrect, having been re-cut at a later date, with the stonemason mistaking the weathered numeral eight for a three.

Today the building once again functions as a roadside inn, with access to the site from the dual carriageway A1 having been restored in the 1980s. In addition there are nineteen bedrooms along with facilities for weddings and conferences.

35 Wansford Great North Road Bridge
Mass concrete road bridge spanning the River Nene
Year of completion: 1929
Structural Engineer/Architect: Sir Owen Williams / Consulting
architects: Maxwell Ayrton and Sir John Simpson
Listed status: Grade II*
Location: 52.581539, -0.412983

Upon its creation the newly designated A1 was littered with potential
bottlenecks as it passed through towns and villages unable to cope
with ever-increasing levels of motorised traffic. With its narrow
sixteenth-century bridge, tight corners and dangerous crossroads,
Wansford was a prime candidate for a relief road. A new route east of
the village was decided upon, spanning the River Nene. Design of the
bridge was entrusted to one of the foremost engineers of the age. In
collaboration with Maxwell Ayrton, 'Concrete Williams' had already
designed the new Empire Stadium at Wembley, as well as a series
of bridges along the route of the A9 in Scotland. These included a
wonderful faceted example at Crubenmore and an unusual concrete
'suspension' bridge at Montrose (since demolished).

This is a mass concrete construction, i.e. it is built without steel
reinforcement. Spans of the three arches are 50 feet, 109 feet and
50 feet. Its appearance was dictated to an extent by the requirement
to blend in with the nearby village. Hence the Romanesque shape of
the arches within the spandrels, from which project wide expansion
joints. There is an impression of weight and strength, though the

44

structure as a whole fails to demonstrate the dynamic, spectacular results that could be achieved with concrete. For an example of this we must look to Robert Maillart and his breathtaking Salginatobel Bridge of 1930.

As with most of Williams' structures from the period, the concrete has weathered from its original dazzling off-white to a dull brown colour. Nowadays it only carries northbound traffic, having been supplemented in 1975 by a new bridge as the Nene crossing was upgraded to dual carriageway.

㊱ Former Wansford Knight
Derelict 'roadhouse' built in moderne style
Year of completion: 1932
Architects: Davies & Knight
Location: 52.582659, -0.413850

This well-known A1 landmark occupies a prominent position on a rise a short distance from Wansford's *Great North Road Bridge* [35]. Many people will remember it best as a branch of Little Chef, but it began life as one of the five 'Knights of the Road', the creation of Mr C. G. Knight. These were a chain of outlets offering refreshments and overnight accommodation for the rapidly expanding motoring classes and were situated at strategic points along the arterial road network of the Midlands (in this case the new Wansford bypass). All were completed in the modish streamlined moderne style, with

rendered brickwork painted white, contrasting green metal-framed casements and flat roofs. Other roadhouses from this era generally stuck to a debased neo-Tudor style, however inappropriate this might now seem for a new age of high-speed, long-distance motoring. Both the Nottingham and Hinckley Knights have since had their pure lines sullied by the addition of incongruous pitched roofs.

The Wansford Knight proved to be a short-lived addition to the assortment of roadside refuges dotted along the A1, being reincarnated four years later as The New Mermaid, this acting as a replacement for an inn which had stood in the centre of Wansford village that was demolished to make way for an improved road junction. Between 1979 and 2007 a franchise of Little Chef operated from the building, providing hot meals for northbound motorists (as well as the much-prized free road map). Along with the Markham Moor diner (see entry 45) this was one of the chain's few architecturally noteworthy outlets, though various alterations made during this period resulted in a later application for listed status being turned down. Abandoned and left to the mercy of the elements and graffiti artists, the structure seemed destined for dereliction and ultimately demolition. Happily, the site has recently been purchased by Harris McCormack Architects, who intend to restore the building and convert it for use as their own offices.

37 Harrier Gate Guardian
Preserved jet aircraft at entrance to RAF Wittering
Year of manufacture: 1990
Design: Hawker Siddeley / Bristol Siddeley
Location: 52.615414, -0.446615

Now that the famous derelict English Electric Lightning XN728 has finally been put out of its misery (it stood for years in the centre of a scrapyard near Balderton, Nottinghamshire), the Harrier gate guardian stands alone on the A1 as a reminder of a bygone era of innovative British military aviation design. This particular airframe, ZD469, has been cosmetically restored after being damaged beyond repair following a rocket attack on Kandahar airbase. Known to RAF personnel as 'Christine', ZD469 became infamous among maintenance crews for her uncanny ability to inflict personal injury while routine maintenance was being carried out. She replaced an earlier Harrier GR3 as the Wittering gate guardian in 2011.

46

RAF Wittering dates back to 1916 and has seen a varied collection of fixed-wing types come and go. Spitfires and Hurricanes of 12 Group operated here during the Second World War, while in the Cold War era it was home to Vickers Valiant nuclear bombers. From 1969 until 2010 Wittering was known as the 'Home of the Harrier', with these aircraft becoming a familiar sight to motorists on the A1 as they flew low in the vicinity. Since the retirement of the Harrier fleet the skies above Wittering have fallen silent, with the base itself now given over to the RAF's expeditionary and logistics force.

38 Holy Cross, Great Ponton
Perpendicular parish church and adjacent manor house
www.ellysmanorhouse.com
Listed status: Grade I
Location: 52.863771, -0.627611

This church is dominated by a very tall west tower which dwarfs the three bay nave. It is finished to a very high standard using precisely cut, dressed ashlar and is lavishly decorated with finely carved friezes at the base and to the parapet. Such workmanship would have been prohibitively expensive and could only have been made possible by the largesse of a wealthy benefactor, in this case one Anthony Ellys (or Ellis), wool merchant of the Staple of Calais. Still under English control in the early sixteenth century, Calais was a major centre for the export of wool to mainland Europe. Those granted the right to trade in this valuable commodity were assured of handsome rewards to take back across the Channel, where they could immortalise themselves by funding improvements to their local church. In case anyone remained in any doubt as to who they should be grateful to for this munificence, the Ellys coat of arms is prominently displayed on the tower.

All of this reconstruction work was based upon an existing structure; sections of thirteenth-century fabric including the porch and part of the chancel arch have survived, as have elements of what are believed to be the tomb of Anthony Ellys in the north chapel and south aisle. A short walk from the church is the manor house built for the same individual, again an enlargement of an earlier building. This has a Flemish crow-stepped gable to the street elevation and, on the first floor, a remarkable survival: a well-preserved series of early sixteenth-century wall paintings depicting woodland scenes.

39 L6 transmission tower
Tapering lattice steel pylon carrying electricity cables
www.pylons.org
Year of completion: 1967
Design & construction: Balfour Beatty & Co. Ltd
Location: 53.013992, -0.764839

In its own way just as successful and enduring a design as Giles Gilbert Scott's classic red telephone kiosk (see entry 12), only held in considerably less affection by the majority of the general public. This despite the invaluable role they play in helping to bring light and heat into our homes. The basic pylon form dates back to the creation of the Central Electricity Board in 1926, and with it the National Grid. A standardised structure for carrying the high voltage cables needed to be agreed upon, with the Board enlisting architect Sir Reginald Blomfield (a vociferous critic of the modern movement in architecture) to give advice on aesthetic considerations. Eventually it was decided to adapt a design created by the American Milliken brothers, with the first examples being erected near Edinburgh in 1928.

There are now 22,000 towers supporting the National Grid's main transmission network. This example, 4VK 90, stands next to the A1 north of the village of Long Bennington. It is part of a line of L6s that run from Cottam power station to a point near Stevenage. They can be seen crossing the A1 at several points as both routes head south, though these 50-metre-high giants have become such a familiar sight that most travellers barely register their existence.

A desire to develop a new tower that would meet the ever-increasing demands of the twenty-first century led to the organisation of a competition, this being won by the Danish firm Bystrup with their 'T'-Pylon. Overcoming some more striking proposals, this simple design promises to be lower, lighter and easier to assemble than the established lattice structure. However, converting the entire network to the new design would be prohibitively expensive, so the steel skeletons will be striding across our landscape for many years to come.

40 Balderton Psychiatric Hospital water tower
Disused art deco water tower within residential redevelopment
Year of completion: *c.* 1940
Architect: The County Architect, Nottinghamshire / Building Contractor: Ernest Coleman Ltd
Location: 53.048358, -0.771276

Now stranded within a genuine mock-Georgian housing estate, this handsome brick tower was once the most prominent element of Balderton Hospital, a 'colony for mental defectives' to use the charming phraseology of the 1930s. This new development was based around Balderton Hall, which dates from 1840 and was used as the hospital's administration block. Originally budgeted at £195,400, work was delayed firstly by the Second World War, then by post-war austerity. It was not until 1961 that Minster for Health Enoch Powell was able to cut the ribbon at the completed site. It survived until 1993, with the water tower and hall now

49

the only surviving elements. All other structures were cleared to make way for Fernwood village.

Originally flanked by matching L-shaped wings (the outline of the gables can still be seen near to the base), the water tower was completed in a simplified art deco style; the uppermost storey tapers and is capped by a stone parapet. Below this all four sides feature a clock face and a slender vertical strip of window to light the interior.

Intended to become an asset to the village in the form of a bar/restaurant, with its comparatively small footprint and lack of natural light the tower would make a dauntingly difficult prospect for conversion. At the time of writing it remains disused and fenced off.

41 All Saints, Winthorpe
Gothic revival parish church in red brick
Year of completion: 1888
Architect: Sidney Gambier-Parry
Listed status: Grade II
Location: 53.098248, -0.788669

This church stands on what was once the main Gainsborough–Newark road, which was severed following the completion of the Newark by-pass in 1964. It is a replacement for an earlier building which was itself substantially re-built in the eighteenth century but deemed too dilapidated to repair. With sturdy buttressing to tower, nave and chancel, All Saints exudes Victorian muscularity, its prominence as a local landmark assured by the use of vivid red brick throughout (with contrasting Ancaster stone dressings) and the tall steeple with a broach spire. This is offset to the north-west, moved from its usual location to make way for a small apsidal baptistery. Currently this contains the font, though there are plans to move this to the nave so the baptistery area can be adapted to contain a disabled toilet and small servery. Designed by CBP Architects, this conversion will be screened from the nave by the addition of a solid timber reredos.

㊷ Winthorpe Bridge
Reinforced concrete bridge over the River Trent
Year of completion: 1964
Engineer: Alfred Goldstein CBE of R. Travers Morgan & Partners
Listed status: Grade II*
Location: 53.101666, -0.798724

With its very wide central span, Winthorpe Bridge appears to defy gravity in much the same way that the railway bridge at Maidenhead appeared to do over a century before. However, by utilising prestressed concrete it was possible to construct a span twice the width that Brunel was able to achieve with brick. It was built as part of the 6½-mile Newark by-pass scheme of the early 1960s and represented the principle engineering challenge of the project. Composed of nine box girders

which were cast on site, the bridge consists of three spans of 130 feet, 260 feet and 130 feet, giving a balanced, symmetrical appearance. Care was taken with the aesthetic qualities of the structure, with different elements being given contrasting facings: smooth concrete for the piers, a vertical fluted effect on the main fascia, topped with a band of blue shap granite aggregate below the parapet.

Viewing the structure from the riverside reveals its impressive scale, daring cantilever construction and harmonious integration with the landscape. English Heritage recognised these qualities by listing it Grade 2* in 1998. It became one of only a handful of post-war bridges or viaducts to be so recognised (see also entry 49). A shame, then, that its isolated location means there are usually only a few anglers around to admire it.

43 St Mary, Carlton-on-Trent
Gothic revival parish church with prominent spire
Year of completion: 1851
Architect: G. G. Place
Listed status: Grade II*
Location: 53.166571, -0.805618

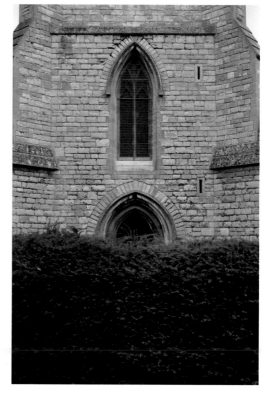

With its slender spire soaring above the low-lying east Nottinghamshire landscape, St Mary's dominates the small village of Carlton-on-Trent. It is typically Victorian in its unity of style, other than the slightly later 'decorated' spire, the whole constructed in attractive honey-coloured ashlar masonry. There is a strong sense of verticality to the structure thanks to the steep-sided nave and chancel roofs and the deep lancets of the west tower. St Mary's is a replacement for an earlier chapel, the most notable remnant of which is the twelfth-century south doorway.

A tombstone in the graveyard bears the following inscription: 'In affectionate remembrance of George Gordon Place, architect of this church, who departed this life the XXVIIIth day of March in the year MDCCCLXXXIX [28 March 1889].'

㊹ Tuxford Windmill
Tower windmill restored to full working order
www.tuxford-windmill.co.uk
Year of completion: *c.* 1810–20
Listed status: Grade II
Location: 53.238072, -0.902762

One of only three windmills remaining in Nottinghamshire still capable of grinding grain (the others are at North Leverton and Sneinton in Nottingham), Tuxford mill was restored during the 1980s and is now run as a business alongside the adjacent café and gift shop. Organic flour ground at the mill can be purchased here, and visitors can also pay a small fee to ascend the tower, via very steep ladders. From here you may glimpse another windmill across the field, this example having lost its cap and sails and been converted into a house.

With its squat, tarred brick tower, three floors, an elegant ogee-shaped cap, eight-bladed fantail and four shuttered 'patent' sails, it features many characteristics that were prevalent in the region during the early nineteenth century. Head further east into Lincolnshire and towers become much taller, with five, six or even eight sails.

Though it now stands on its own, the existing building was once one of a pair of windmills on the site. Surviving photographs show a wooden 'post' mill with a very rare twin fantail arrangement attached to the rear of the buck. This technical novelty did not prevent the structure from being demolished in the 1930s.

㊺ Former petrol station, Markham Moor
Concrete 'hypar' canopy to former filling station
Year of completion: 1960
Architect: H. S. (Sam) Scorer / Structural engineer: Dr K. Hajnal-Kónyi
Listed status: Grade II
Location: 53.256899, -0.927730

That most humble and utilitarian of buildings, the petrol filling station, is here given a spectacular flourish with the addition of a saddle-shaped

concrete 'hypar' canopy. The result of a collaboration between Scorer and Hungarian émigré Hajnal-Kónyi, Markham Moor is a rare surviving example of a hyperbolic paraboloid roof, an engineering solution which enjoyed a brief period of popularity in the 1950s and 60s. They are usually double-curved geometric structures supported by columns, leaving a clear unobstructed space underneath. Scorer also penned two other hypars, which are still extant in Lincoln: St John the Baptist parish church and a car showroom (now a restaurant). Both are listed.

To visualise the structure as completed, we must discount the single-storey restaurant constructed underneath in the 1980s, which was built for Little Chef after the petrol station closed down. Despite its futuristic appearance you can't help but wonder quite how effective the roof was in terms of protecting customers from the elements. It did at least function as an unmistakeable beacon for approaching motorists eager to re-fuel their Ford Anglias.

English Heritage finally recognised this as a significant example of transport architecture by listing it Grade II in 2012. Despite this layer of protection the building is currently empty, decaying and facing an uncertain future. Hopefully somebody can adaptively re-use this most distinctive landmark before it deteriorates any further.

46 Priory Church of St Mary and St Martin, Blyth
Parish church with Norman nave and medieval 'doom' wall painting
Year of completion: *c.* 1088
Listed status: Grade I
Location: 53.378806, -1.063396

What we see at Blyth today is a building that has been adapted, extended and remodelled countless times over nearly a millennium. As approached through the graveyard towards the south porch, the dominant features are an Early English (thirteenth-century) south aisle, and a soaring perpendicular west tower. The original structure (part of a Benedictine priory) is not at all apparent. This was largely swept away in the seventeenth century to make way for Blyth Hall, itself now demolished. Today the priory church finds itself the focal point of a modern residential development, made up of large detached houses.

Of the church that was completed near the end of the eleventh century, only a fragment remains. This is the Romanesque nave, with simple round arches and a general sense of solidity and heft that is characteristic of Norman design, here offset somewhat by the later addition of a vaulted ceiling. Of the rest of the interior there is much to admire: stained glass by the firm of C. E. Kempe in the west window, a monument to Edward Mellish (builder of Blyth Hall) and some evocative paintings of saints in the south aisle screen. However the greatest treasure is the restored fifteenth-century 'Doom' painting adorning the east wall, which was only rediscovered and restored in the 1980s. An unknown hand has depicted the Day of Judgement on an epic scale. Despite the inevitable erosion of detail through age and neglect, the mural retains an elemental power.

47 Harworth Colliery winding towers
Pair of headstocks at mothballed coal mine
www.ukcoal.com
Year of completion: 1989 (No. 1 shaft), 1996 (No. 2 shaft)
Consultant architects: Faulks Perry Culley & Rech / Consultant engineers: British Mining Consultants Ltd
Location: 53.414395, -1.060527

A concrete cathedral dedicated to king coal, the 80-metre tower over No. 1 shaft at Harworth dominates the skyline for miles around, with only the similar outlines of the headstocks at nearby Maltby Main Colliery (closed as of 2013) for competition*. Completed at a time when the Thatcher administration was committed to a programme of wholesale pit closures, this major investment offered some hope that there was a future for mining in Britain. However, there are now just three deep mines remaining in the country, with Harworth currently dormant and on a 'care and maintenance' programme.

Originally sunk in the 1920s, Harworth lies within the northern border of Nottinghamshire but can be perhaps considered a South Yorkshire mine due to its position on that region's coalfield. It was a producer of 'hard' coal whose qualities made it suitable for use in locomotives. Celebrated engines of the great age of steam such as *Mallard* and the *Flying Scotsman* were reported to be powered by Harworth or Bentley coal when making their record breaking attempts in the 1930s.

Constructed from slip-formed concrete, No. 1 headstock was built around and above the existing winding tower so as not to disturb production. Once this was complete the redundant older structure was cut through at its base and towed to a position 20 metres to the north – a remarkable engineering achievement which helped gain the project a commendation in the 1990 Civic Trust awards. Seven years later a new 45-metre No. 2 winding tower was inched into place to supplant the obsolete steel and concrete headstock. Its blue and yellow exterior cladding matched that of its larger neighbour and was an attempt to reduce the visual impact of these huge new additions to the landscape.

* Since demolished.

Part Three: The North

48 Robin Hood's Well
Ornamental well cover
Year of completion: *c.* 1710
Architect: Sir John Vanbrugh
Listed status: Grade II
Location: 53.599753, -1.217145

This curiosity sits at the end of a lay-by on the southbound side of the A1. It is a minor work by the renowned architect Sir John Vanbrugh, who had been commissioned to design it for the 3rd Earl of Carlisle. Vanbrugh was already working on the opulent Castle Howard for the same client. Built from magnesian limestone ashlar, this ornamental structure originally covered an artesian spring from which pure drinking water was obtained. For centuries, passing travellers would pause to replenish water supplies using the chained ladle provided, before continuing with their journey along the old Great North Road.

Widening of the A1 as part of the Redhouse to Wentbridge scheme of 1960/61 necessitated that the well cover be dismantled and moved from its original location. All of the blocks were carefully numbered so that they could be reassembled at a new site to the south of the spring, this time resting on a base of solid concrete. Engravings from the eighteenth and nineteenth centuries appear to suggest that the structure was originally at least one course higher than it currently stands. Possibly this truncation took place at the time of its relocation.

49 Wentbridge Viaduct

Prestressed concrete viaduct over the Went Valley
Year of completion: 1961
Senior designer and engineer: F. A. Sims
Listed status: Grade II
Location: 53.648982, -1.255345

When approached from the public footpath leading from Wentbridge village, this vast structure demonstrates how bold, innovative civil engineering can harmonise with even the most picturesque landscape. Both the Royal Fine Art Commission (who were consulted at the planning stage) and the Museum of Modern Art in New York acknowledged the aesthetic qualities of this viaduct, the latter including it in their exhibition of twentieth-century engineering in June 1964. By 1998 it had become one of the few post-war spans to be listed, along with its near contemporary *Winthorpe Bridge* [42].

Despite appearances, both the abutments and much of the deck is actually hollow, which enabled an innovation whereby the prestressed tendons were placed in the voids of the deck rather than being embedded in the concrete. This strengthening allowed a total span of 144 metres, supported by two pairs of angled piers which bury themselves into the edge of the valley floor. For the seriously intrepid pedestrian, footpaths are cantilevered on either side of the dual carriageway. No attempt has been made to incorporate any decorative elements into the overall design, which is all the more striking and impressive for it.

50 Limestone 'barrows' at Holmfield Interchange
Landscape features inspired by local archaeology
Year of completion: 2006
Design: RPS Consultants*
Location: 53.712141, -1.287104

Motorists taking the A1(M) northbound past Holmfield Interchange may catch a fleeting glimpse of five mysterious conical features, positioned in a flat no-man's land where the A1(M) intersects with the M62. These artificial 'barrows' or tumuli are a result of a competition organised to create a visual reminder of the extensive archaeology located along the route of the Darrington to Dishforth motorway upgrade. Probably the most notable of these are Ferrybridge Henge and an Iron Age chariot burial.

In the shadow of *Ferrybridge C power station* [51] is the site of a Neolithic henge, a circular earthwork where religious ceremonies would have taken place. By the end of the Bronze Age several burial mounds had also been completed in the vicinity. Very little of this prehistoric activity is now apparent above ground, meaning that this particular henge has remained largely unknown to the general public. However, a more recent discovery at Ferry Fryston received worldwide media coverage. Excavation of a square barrow revealed the remains of an Iron Age chariot which had been buried in a complete state, an extremely rare occurrence indicating that this was the grave of a man of considerable status. His tomb would originally have been covered by a mound formed from the excavated limestone, creating

a prominent landmark and inspiring the features we see today at the new motorway junction.

*I have been unable to find out the name of the design competition winner but would be delighted to include this information in an amended version of this book.

51 Ferrybridge 'C' power station
Coal-fired power station
www.sse.com/ferrybridge
Year of completion: 1966 (electricity first fed into the national grid)
Location: 53.716038, -1.280392

This vast industrial complex announces its presence long before the A1(M) skirts its western boundary at Holmfield Interchange. Motorists will first notice a cloud of condensed water vapour rising lazily into the sky, then the cooling towers and chimneys break the horizon. These are giants, the tallest structures encountered anywhere on our journey from London to Edinburgh. Eight closely grouped cooling towers rise to 115 metres, with the two main chimneys soaring to 198 metres. They are the most prominent elements of a facility that was first opened in 1924. As its name suggests, this is its third incarnation. Surviving within the perimeter of the current station is the turbine hall of Ferrybridge A, a Grade II listed brick building with a classical façade which is now used as offices and workshops. Ferrybridge B operated alongside C for twenty-six years before being decommissioned in 1992 and subsequently demolished.

With the environmental impact of coal-fired power stations apparent, current owners Scottish & Southern Energy are attempting to secure the facility's long-term future by investing in new technology. Two of the four 490 MW generator units have been fitted with Flue-gas Desulphurisation technology and a 68 MW multifuel generation plant is due to open in 2015. Using coal as a viable future power source is dependent upon the development of CCS (Carbon Capture and Storage) systems. To this end SSE has pressed ahead with a pilot scheme at Ferrybridge, which is capable of capturing a small percentage of current carbon dioxide emissions.

52 Moto Wetherby MSA
Motorway service area designed for maximum energy efficiency
Year of completion: 2008
Architects: JWA Architects
Location: 53.946841, -1.368732

This striking building gives an indication that MSA (Motorway Service Area) operators have realised that attractive, considered designs may help to lure motorists off the motorway and into their concessions. In many ways it is a return to the earliest days of the MSA in the 1960s, when developers created prominent landmark buildings for a public eager to experience the new motorway network. Perhaps the most celebrated of these were the Pennine Tower at Forton Services on the M6, which evoked the design of an aircraft control tower (now listed Grade II) and Leicester Forest East on the M1, whose restaurants were

located on a wide bridge crossing the carriageways. In both instances the idea was that diners could gaze down on the spectacle of traffic passing below at unprecedented speeds of 70 mph.

Such enthusiasm for the visual delights of the motorway environment is hard to imagine in the modern era of congestion and tailbacks. Instead the emphasis is on providing respite from the rigours of twenty-first-century travel. At Wetherby another step forward has been taken in the consideration of the environmental impact of these facilities. To this end natural ventilation, solar panel arrays and ground source heat pumps have all been utilised to try to reduce the building's carbon footprint. New areas of woodland have been created along with swales which it is hoped will encourage amphibian wildlife.

The main building is entered via a rotunda which leads customers into the main concourse, a fully glazed seating area with an angled front elevation. It is capped by a bladelike oversailing roof finished in Western Red Cedar supported by irregularly angled columns. There seems to be more than a little inspiration drawn from Zaha Hadid here, in particular her Vitra Fire Station (1993). A petrol station canopy of similar design is located near the exit of the complex.

53 Temple of Victory
Restored Georgian hilltop folly
www.allertoncastle.co.uk
Year of completion: *c.* 1770
Architect: attributed to James Paine
Listed status: Grade II*
Location: 54.020092, -1.377808

Several conflicting accounts exist regarding the commissioning and construction of this picturesque landmark, which occupies the highest point of Allerton Park. It seems most likely to have been built *c.* 1770 to the designs of James Paine for the 2nd Viscount Galway, owner of the estate at that time. The victory commemorated by this temple was probably the Seven Years' War of 1756–63; Robert Adam was completing a similar memorial at Audley End during this period. Replacing an arbour, the Temple is octagonal with a domed roof and was used as a banqueting house. It was also illuminated at night to denote British military victories such as Trafalgar (1805).

All but abandoned and left to decay, by the 1980s the Temple was in a ruinous condition. Extensively vandalised, the interior was also left

open to the elements due to the theft of most of the lead-covered roof. It was only saved by the intervention of the new owner of Allerton Castle, Dr Gerald Rolph. Holland Brown Architects were commissioned to restore the external appearance of the building. This involved the reconstruction of the roof and repairs to the ashlar fabric. Rather than replace the vulnerable glass, it was decided to cover the window openings with sheet material, which was then painted to imitate the original sash windows. Viewed from a distance the effect is entirely convincing.

54 Angel of the North
Landmark sculpture on reclaimed industrial site
www.gateshead.gov.uk/angel/
Year of completion: 1998
Designer: Antony Gormley / Consulting Engineers: Arup / Manufacture: Hartlepool Steel Fabrications
Location: 54.914155, -1.589519

'People are always asking why an angel? The only response I can give is that no-one has ever seen one and we need to keep imagining them.' (Antony Gormley)

Standing impassively above the Team Valley at the gateway to the Tyneside conurbation, this is an angel that is destined to forever remain earthbound. Seemingly poised to take flight, it is anchored to solid rock 20 metres below ground level in order to withstand

63

100 mph winds. A strange hybrid of a recognisably human form married to wings that seem to belong more to the modern era of aeronautics, clad in a thick armour of weathered cor-ten steel. One of the most recent landmarks on this list, the *Angel of the North* is without doubt now the most famous. Indeed, it is now hard to recall a time before it existed.

There would have been no Angel had it not been for the determination of Gateshead Council to promote public art through its own programme, beginning in 1986. With no dedicated space for displaying artworks at that time, the town itself became the backdrop for a series of murals, installations and sculptures. The programme was well established by the time a competition was held for a landmark sculpture on the site of the former Team Colliery. Impressed by his ability to engage with local communities (as demonstrated in his remarkable *Field for the British Isles*) and no doubt seeing the potential in *A Case for an Angel* (an earlier, smaller winged figure), Gormley's designs were selected by the panel. Total cost was £800,000, all of it being met through grants and donations rather than from local council taxes.

Fears about the Angel being a distraction to motorists on the A1 have proved to be unfounded, with no major accidents attributable to the sculpture. It is believed that as it becomes visible to drivers from a distance, they have time to absorb its presence gradually, rather than being startled.

55 Northumberlandia
Landform sculpture of reclining female
www.northumberlandia.com
Year of completion: 2012
Designer: Charles Jencks
Location: 55.087968, -1.629559

Not a prehistoric creation representing a pagan goddess but a twenty-first-century attempt to create a community park containing an iconic landmark feature, using material extracted from the adjacent Shotton opencast mine. It was co-funded by land owners the Blagdon Estate and mine operators the Banks Group, who commissioned the landscape architect and theorist Charles Jencks to design a 'landform' that may in time become a potent symbol of the county from which it takes its name. Presumably the success of the *Angel of the North* [54] as a tourist attraction provided a spur to development too, though conceptually it is probably closer to Northala Fields in Western London: both use reclaimed spoil (in the case of Northala Fields, concrete from the demolished old Wembley Stadium) to create a viewing platform next to a major road.

Members of the public gain entry via a car park on Blagdon Lane. A path leads through a small plantation, at the exit of which *Northumberlandia* reveals her profile, with the facial features of this giantess immediately apparent. Visitors can ascend a series of steep paths which culminate in a windswept observation point atop the forehead. From here onlookers are rewarded with spectacular views of the surrounding countryside, including the Cheviot Hills, from which Jencks drew inspiration when designing this dramatic new addition to the landscape.

Pair of weatherboarded neoclassical bus shelters
Year of completion: 1937
Architect: Sir Laurence Whistler
Listed status: Grade II
Location: 55.108287, -1.666303

Anyone waiting for the number 44 from Stannington to Morpeth can
shelter from the elements in some style in this, one of an identical
pair of bus shelters presented to the villagers by Lord Ridley,
owner of nearby Blagdon Estate, within which *Northumberlandia*
[55] has recently been created. They were built to commemorate
the coronation of George VI and are believed to be the work of
Sir Laurence Whistler, later to become famous for reviving the
art of glass engraving. Entry is via a doorway framed by pilasters
with a date-inscribed lintel above, while the interior is lit by a
Georgian-style window.

57 Nelson Monument
Commemorative stone obelisk
Year of completion: 1807
Mason: Thomas Robson
Listed status: Grade II
Location: 55.319943, -1.726618

One of a multitude of memorials dedicated to the memory of Lord Nelson following his death at the Battle of Trafalgar (see also *The Nelson Monument* [75]), this masonry obelisk was also specifically intended to mark the 'memory of private friendship' between Nelson and Alexander Davison, a controversial figure who would serve two separate prison sentences, one for bribing voters in elections and another for defrauding His Majesty's government in the supply of stores for military barracks. Davison was then the owner of nearby Swarland Hall (long since demolished), in the grounds of which he had arranged for trees to be planted in such a configuration that, when seen from above, they would represent the coastline and the positions of Nelson's fleet during a key moment of the Battle of the Nile.

After becoming neglected and overgrown, the monument has recently been restored with the aid of local council and Heritage Lottery Fund grants.

58 Limekiln
Disused nineteenth-century sandstone structure used for producing quicklime
Listed status: Grade II
Location: 55.576540, -1.797525

Northumberland's abundant limestone deposits have for centuries been exploited for use in the building trade and in agriculture. Remnants of this industry can be found across the county, often in rather isolated locations such as this example on the narrow lane down to the hamlet of Adderstone Mains. Long since abandoned and partially overgrown, it is quite difficult to see how this unit would have functioned when in operation. A better preserved

example can be found at Great Tosson. Here we can see the artificial mound from which the limestone and a burning agent (usually some form of coal) could be more easily loaded into the cylindrical 'pot', with the access arches containing the draw-holes at the base of the structure. A fire would be lit underneath this material which would be allowed to burn at a very high temperature for several days, eventually reducing the limestone to a fine powder – calcium oxide, or quicklime as it is more commonly known. This, when cooled, could then be spread onto neighbouring fields at a ratio of 4 tons to the acre.

This inefficient method of producing quicklime has long since been superseded by modern machinery which can easily pulverise limestone, while in agriculture the development of chemical fertilisers was also a factor in the decline of the traditional kilns. Those that survive serve as a reminder of a bygone phase of our industrial development.

59 Haggerston Castle
Surviving remnants of baroque mansion within holiday park
Date of completion: 1897
Architect: Richard Norman Shaw
Listed status: Grade II
Location: 55.686486, -1.935043

Set within the sprawling Haven holiday park are a few fragmented reminders of this site's former existence as the estate of Mr C. J. Leyland,

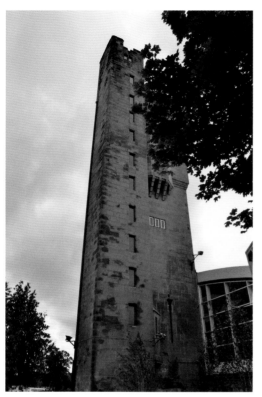

remembered today mainly for giving his name to the quick-growing hybrid cypress (*C. Leylandii*) that has caused so many neighbourhood disputes. Most prominent is the twelve-storey tower that housed the estate's water supply and was also used as an observatory by Leyland. Another survival is the truncated entrance rotunda which gives an indication of the scale of the mansion, an enlargement of an earlier house that was completed in the restrained classical style favoured by Shaw at this point in his career. Prior to this he was a student of the Gothic Revival and of the English vernacular tradition. His most celebrated work is probably Cragside, another Northumberland mansion which was built for Lord Armstrong on a dramatic hillside setting, packed with the latest technology of the day and finished in a distinctive neo-Tudor style.

On Leyland's death the estate, ruinously expensive to maintain, was sold by his son, with the house being stripped of its expensive fittings before being largely demolished. Its footprint is now roughly occupied by the entertainment complex of the holiday park. Though holidaymakers might guess that the castellated tower is an original section of a medieval stronghold, there are no remaining traces of the fourteenth-century 'strong tower' built by Robertus de Hagerston, who had been given a royal licence to crenellate in 1345. This fortification survived as a ruin until 1805.

60 Windpump
Disused windpump of American farm type
Location: 55.708553, -1.971298

Though usually associated with the development of the American West, these multi-bladed machines were once also a familiar sight in Britain, but they are now vanishing almost unnoticed from the landscape. Many of those indicated on Ordnance Survey maps have disappeared or are in an advanced state of decay. This example is complete, though the governor spring that allows the rotor to 'feather'

(turn out of the wind) in potentially damaging gales appears to be broken, meaning that it is no longer capable of facing into the wind.

These windpumps were normally used to raise water from wells to sustain livestock or to irrigate land. This would generally be stored in an adjacent tank. A simple but ingenious design meant that these structures were reliable, self-regulating and low-maintenance (usually just a change of gearbox oil each year was required). They were also capable of operating for many decades. This basic design is still popular in more remote rural areas of the developing world where electricity supplies are unavailable or unreliable.

61 Scremerston Colliery
Engine houses from defunct nineteenth-century coal mine
Year of completion: 1840
Engineer: Thomas Forster / Builder: W. Elliot
Listed status: Grade II
Location: 55.729489, -1.985729

Hidden in a small wood at the southern end of the Berwick bypass are two very sturdily constructed engine houses, reminders of Northumberland's once extensive coal mining industry. This was the Jack Tar or Greenwich pit, the latter name indicating that this

was a venture of the Commissioners of Greenwich Hospital, who had inherited significant tracts of land following the suppression of the Jacobite rising of 1715. It would have been one of several pits in the area and was connected to the nearby Restoration Colliery by a waggonway, the path of which is still obvious today as a wooded cutting heading north. Coal from Scremerston would have been transported by rail to Berwick-upon-Tweed via Spittal. The Cooper Eye seam was worked until 1944.

Of the surviving structures, the pumping-engine house is in better condition, its interior protected from the elements by a rooftop cast-iron water tank. Installed within was a beam-type steam engine that would have pumped water from the coal seams. An inscribed plaque between two round-headed windows on the second floor records the date of completion. Adjacent to this is the winding-engine house, which has lost its roof and is now in an advanced state of deterioration.

Part Four: Scotland

(62) Ayton Castle
Large country house built in Scottish Baronial Style
Year of completion: 1851
Architect: James Gillespie Graham
Listed status: Category A
Location: 55.845616, -2.114928

Situated on high ground overlooking Eye Water, the site of the present-day country house was for centuries a strategically important strongpoint, heavily fortified against incursions from the south. Captured by forces commanded by the Earl of Surrey in 1497, the castle was described as being 'very sore dismantled' by 1544 and no trace remains of it today. From this period, only the ruin of the twelfth-century church of St Dionysius and a sixteenth-century dovecote remain. Eventually the estate passed into the hands of the Fordyce family, who constructed a classical mansion, only for this to be consumed by fire in 1834. It was the destruction of this building which allowed the subsequent owner, William Mitchell-Innes (a director of the Royal Bank of Scotland), to commission a new structure in a fashionable revivalist style.

With its four-storey tower that resembles a keep, crenellated parapets, crow-stepped gables, corner turrets with conical roofs and arrow slits, Ayton Castle is a defining example of the Scottish Baronial style. This commission came near the end of a long and distinguished

career for Gillespie Graham, who was versatile enough to also produce neoclassical and Gothic Revival designs elsewhere in Scotland. There are seventeen bedrooms and nine bathrooms within an opulent interior decorated by the firm of Bonnar & Carfrae. Besides the castle itself, Gillespie Graham also added a fine south lodge in matching style and planned the rebuilding of Ayton village (bypassed by the A1 in 1981).

(63) Dunglass Viaduct
Masonry viaduct carrying railway across gorge
Year of completion: 1846
Engineer: John Miller
Listed status: Category A
Location: 55.941881, -2.368785

One of a quintet of bridges that cross the gorge of the Dunglass Burn north of Cockburnspath on a well-worn route to Edinburgh that has been used for centuries, albeit with various alignments at this point. Four of the bridges were built for road traffic: in order of age they are the seventeenth-century Old Bridge which is nearest to the coast and shrouded by thick undergrowth, the castellated 'New' Bridge of *c.* 1800 that was replaced by the 1930s *Dunglass Road Bridge* [64] and finally a 1990s composite structure that currently carries the A1 across the gorge. Towering over all of these is bridge ECM8/109 (as it is evocatively titled by Network Rail), which is now a section of the East Coast Main Line but originally formed part of the North British

Railway's ambitious 57-mile Edinburgh to Berwick line. Miller was chief engineer for this scheme, which represented just one project in a hugely prolific career that coincided with the period of 'railway mania' across the country.

Composed of six semicircular arches and with a principal span of 41 metres, this viaduct impresses with its scale rather than with any elaborate decoration. It is faced in sandstone that has been dressed on the flanking piers and the radial voussoirs. Other than the addition of overhead power lines when the ECML was electrified in the late 1980s, the structure has survived virtually unaltered.

(64) Dunglass Road Bridge
Reinforced concrete arch bridge spanning deep gorge
Year of completion: 1932
Engineers: Blyth & Blyth
Listed status: Category B
Location: 55.942244, -2.368196

Like an oxbow lake cut off from a river, this bridge has been severed from the main road network but survives to indicate one of the previous paths taken by the A1 over the steep-sided gorge of the Dunglass Burn. It replaced the impracticably narrow 'New' Bridge (*c.* 1800), which still stands about 100 metres upstream. Engineered by the same company that constructed Edinburgh's North Bridge (who are still in existence today), this handsome open-spandrel

design is composed of five mighty concrete ribs, with the road deck supported by a series of arches on columns. Central span is 48 metres. It is instructive to compare Blyth & Blyth's engineering solutions to those employed by Owen Williams at his *Wansford Great North Road Bridge* [35], another significant concrete A1 crossing from the same era.

Deemed too weak to withstand modern traffic levels and apparently earmarked for demolition at one point, the bridge was superseded as part of the Tower to Dunglass A1 improvements, which bypassed the village of Cockburnspath. Its replacement, disappointingly utilitarian in such a picturesque setting, is a composite structure supported by two giant concrete finger piers.

65 Torness nuclear power station
2nd generation nuclear power station with gas-cooled reactors
www.edfenergy.com/energy/power-stations/torness
Year of completion: 1988
Design & Construction: Sir Robert McAlpine / Reactor Design & Engineering: NNC
Location: 55.967413, -2.410048

For the last quarter of a century, Torness Point has been dominated by the slightly sinister form of this 1,190 MW power station, the tall blank walls concealing two advanced gas-cooled reactors that are capable of supplying electricity to over 2 million homes in the UK. It is one of

two remaining nuclear power plants in Scotland (Hunterston B in North Ayrshire being the other) and is expected to remain operational until at least 2023. Locals and travellers on the A1 alike have become used to its monolithic presence, though there was fierce opposition to its construction, with protestors occupying the site in largely peaceful demonstrations that drew nationwide attention to this corner of East Lothian. Robin Cook, then Member of Parliament for Edinburgh Central and future Foreign Secretary, was among the campaigners.

This type of facility's safety record will inevitably be carefully scrutinised, though thankfully there have only been a handful of noteworthy incidents. Of these, two stand out as being particularly unusual. In June 2011 both reactors had to be manually shut down when it was discovered that the water cooling system filters were being obstructed by huge amounts of jellyfish. Potentially far more serious was the total loss of an RAF Tornado F3 during November 1999. On a low-level training flight, the aircraft crashed into the sea 13.5 kilometres from Torness Point following engine failure. Its crew, both of whom survived, were subsequently praised for their actions in ensuring the aircraft was pointing out to sea and away from the power station before ejecting. A potentially catastrophic incident was therefore avoided.

66 Tyne Bridge
Post-tensioned concrete road bridge
Year of completion: 2004
Design: Scott Wilson Scotland / Engineering: Balfour Beatty
Location: 55.977390, -2.666908

Is it possible for a large-scale engineering project to complement or even enhance an existing area of outstanding natural beauty? This was the question facing the Scottish Executive during the planning stages of the £35 million, 13.7-km Haddington to Dunbar 'expressway' that would cut right across the attractive River Tyne valley. After consultation with the Royal Fine Art Commission for Scotland and Scottish Natural Heritage, it was decided that extra funding would be provided in order to ensure an aesthetically pleasing design. What emerged was a post-tensioned concrete structure built on 45-degree splayed legs that crosses the valley in a gentle curve.

Opting for this type of bridge would lead to many exacting engineering challenges for the Balfour Beatty team. During construction the entire structure was supported by a vast, intricate network of

steel formwork on a scale that had no precedent in Scottish history. Another ambitious element was the sixteen tensioning tendons that were designed to run the entire 217-metre length of the bridge, adding strength and reducing visual bulk. In order to ensure a successful installation, post tensioning specialists Balvac completed a full-scale trial which demonstrated the feasibility of the system.

This careful and considered approach to a potentially damaging and intrusive river crossing has resulted in an elegant structure, one which helped the entire expressway project be shortlisted for the Prime Minister's Award for Better Public Building in 2004.

67 Dolphingstone Doocot
Seventeenth-century stone tower built to house pigeons
Listed status: Category A
Location: 55.943858, -2.991461

Along with Fife, East Lothian has the greatest remaining concentration of doocots (dovecote for those south of the border) remaining in Scotland. This is a fairly typical, if poorly preserved, example of the earlier type, being a tapering circular tower with a central oculus in the roof to permit the entry and exit of the pigeons. Originally the internal walls would have been lined with nesting boxes from which the squabs (the young flightless birds) would have been plucked with the aid of a potence – a revolving timber post with wooden arms from which ladders could be attached so that all the boxes might be reached.

Through the medieval period pigeon was a rare and expensive delicacy for the wealthy few, on whose land the doocots were normally constructed. This example is linked with Cowthrople House which once stood nearby but of which all traces have now vanished.

Gradually these beehive-shaped doocots were superseded by larger rectangular plan designs, which featured a south-facing monopitch roof on which the birds could perch. Decorative exterior features also began to be incorporated to reflect the prestige and affluence of their owners. West Lothian's doocots survive in varying states of repair, with the one at Dolphingstone being on the Buildings at Risk Register for Scotland. In a far better state of preservation is Phantassie doocot near to East Linton, which sports an unusual horseshoe parapet that protects a sloping tiled roof.

68 Queen Margaret University
Purpose-built university campus
www.qmu.ac.uk
Year of completion: 2007
Architects: Dyer Associates
Location: 55.930686, -3.073920

A relatively new university, albeit one that can trace its roots back to 1875 and the formation of the Edinburgh School of Cookery. A series of incorporations of other schools and colleges followed, with the resultant institution (by this point known as the Edinburgh College of Domestic Science) occupying a purpose-built campus at Clermiston in 1970. However, as the college continued to broaden its range of courses into the wider field of the health sciences another move, this time beyond the beyond the boundaries of Edinburgh, became necessary. The site at Clermiston was vacated in 2007 and is now a housing estate. In its place an entirely new campus was constructed on a greenfield site close to Musselburgh, with good road and rail links with Edinburgh.

Awards and nominations for awards were soon bestowed on the development, some more welcome than others. Certainly its

environmental credentials were impeccable, with the Concrete Centre declaring it the winner of their sustainability award for 2006. It also gained a BREEAM* rating of 'excellent' and was shortlisted for the Edinburgh Architectural Awards Building of the Year 2008. Elsewhere, the critics sharpened their knives. *Building Design* magazine awarded the new campus third place in its hotly contested Carbuncle Cup awards for 2009, describing it as 'a building of such depressing PFI drabness it needs to be slapped down'. Presumably they were dismayed by the extensive use of composite cladding and irregular fenestration, those unmistakable hallmarks of building projects nationwide that seek to create 'distinctive' elevations within a tight budget.

Far more satisfying are a pair of equine sculptures on loan to QMU for the winter of 2014/15. *The Kelpies* are scale models of the 30-metre-high stainless steel horse heads designed by Andy Scott which form the centrepiece to The Helix project near Falkirk.

*Building Research Establishment's Environmental Assessment Method.

69 Meadowbank Sports Centre
Amenity for competitive and recreational sports
Year of completion: 1970
Architects: S. L. Harris, T. R. Hughes and B. T. Armistead of the City Architects Department
Location: 55.956775, -3.156334

This description must serve to document a sporting venue whose time has almost passed in its current configuration. As of 2015 plans are in place to replace all the current buildings with a completely rebuilt £43 million centre which may open as soon as 2018. It is to be part-funded by the sale of around half of the 10.1-hectare site for residential development. Fans of track cycling will be disappointed to hear that no replacement velodrome will be included in the proposed scheme, while the 8,320-seat cantilevered main stand will make way for a much smaller structure and therefore mean that national and international athletics events will be a thing of the past.

Construction began belatedly in 1968 on the site of a former speedway venue, following the usual political disputes over funding which seem to be the natural way of things in Edinburgh. The dual ambition was for a venue that could host major athletics events (the Commonwealth Games were held here in 1970 and 1986) as well as serving the needs of the local community with all-weather, year-round sporting facilities. Three indoor halls of varying sizes were provided and survive largely unaltered. These are fairly undistinguished windowless boxes clad in grey steel sheeting, with interior walls composed of striated concrete blocks and a timber v-pattern ceiling. Adjoining these buildings is the main athletics stadium, with the concrete/steel composite stand providing covered accommodation and a shallow bank of seating encircling the rest of the arena. A late addition to the complex was the velodrome which is at the eastern extremity of the site. This is finished in timber with a track surface of Afzelia hardwood strips.

Though hardly ideal for hosting professional football, the centre was formerly home to Meadowbank Thistle, who played here in front of modest crowds from 1974–95 before decamping to Livingston. Non-league Leith Athletic and Edinburgh City remain as tenants.

⑦⓪ Holyrood Abbey Church
United Presbyterian church in 'Free Gothic' style
Year of completion: 1899
Architect: Robert Macfarlane Cameron
Listed status: Category B
Location: 55.957010, -3.164198

Head south from central Edinburgh on the A1 and you will drive past this kirk, prominently situated at the junction of London Road and Marionville Road. It was completed just before the UP Church merged with the Free Church in 1900. Total cost was £6,000. Built in a mixture of grey and red sandstone, its west front features a pair of towers topped by short pyramidal spires flanking a large window with ogee hood mould. The opposite end of the nave is pierced by large transepts to north and south. Cameron was clearly a pragmatist: his list of works from this period seem to be either public houses or religious buildings, a noteworthy example of the former being the Guildford Arms in central Edinburgh, which has a fabulously ornate interior.

A 2007 addition to the site is an entirely new church hall and vestry by Malcolm Fraser Architects, which replaced one dating from 1915. A far more versatile space, it is clad in finely dressed red sandstone

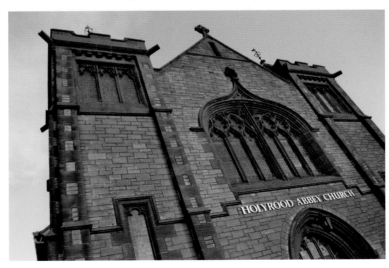

and features extensive glazing, with the large south facing window framing a view of Arthur's Seat. Alterations to the nave of the church were also undertaken at this time, including the replacement of the original pews and the installation of underfloor heating.

As at nearby New Restalrig Parish Church (entry 111), an Evangelical minister and congregation have recently resigned from the Church of Scotland following the Kirk's decision to allow the ordination of gay clergy. They now hold services in Portobello and Leith.

71 Montrose House
Scottish Baronial-style café converted from public house
www.just-the-ticket.biz
Year of completion: c1835
Listed status: Category B
Location: 55.956710, -3.172012

Enlivening the eastern approaches to central Edinburgh is this whimsical Victorian pub front, complete with Scots Baronial turret at the apex of a triangular plan. This rendered façade embellished with ground floor pilasters is believed to be a conversion of an existing building; it appears as a public house on the 1877 Ordnance Survey sheet and another map of 1836, but the plot is vacant on John Wood's detailed plan of 1831, giving an indication of the age of the original structure. It is now a café delivering full cooked breakfasts for £6 (£6.50 if you want scrambled eggs).

Besides the distinctive turret, there are two other noteworthy features on the upper floor of the building. A pair of decorative dormer heads, one employing a semi-circular design apparently based on those at Maybole Castle in Ayrshire, the other bearing a relief of a stag's head with a cross above. This alludes to the fable of King David I and his encounter with such a beast, his deliverance from which by the miraculous appearance of a silver cross inspiring him to found nearby Holyrood (Holy Cross) Abbey.

72 Burns Monument
Grecian commemorative monument to Robert Burns
www.edinburghmuseums.org.uk
Year of completion: 1831
Architect: Thomas Hamilton
Listed status: Category A
Location: 55.953487, -3.178290

While his design for the nearby *Royal High School* [73] was realised in the Doric order, Hamilton settled on more flamboyant Corinthian capitals (externally at least) in fashioning this memorial to the Scottish Bard. It was actually something of an afterthought, built using funds left over from the completion of a marble statue of Burns by John Flaxman. This was initially housed in the new structure but quickly moved when it was realised that the marble was being damaged by air pollution. It now stands in the National Portrait Gallery of Scotland. Hamilton was the natural choice as architect, having already designed a similar memorial to Burns in the poet's home village of Alloway. Both owe their inspiration to the Choragic Monument of Lysicrates, which was depicted by Stuart and Revett in *The Antiquities of Athens*. Constructed from sandstone, the monument takes the form of a cylinder sitting on a rectangular base, ringed by fluted columns. An elaborately carved finial supported by griffons is the crowning flourish.

A gradual deterioration of the condition of the monument, exacerbated by the exposed site on which it stands, necessitated a comprehensive restoration programme which was completed in 2009. As at the Scott Monument, the structure was not cleaned, meaning that the beautifully crafted new sections are readily apparent.

73 New Parliament House (Old Royal High School)
Former school built in Doric Greek Revival style
Year of completion: 1829
Architect: Thomas Hamilton
Listed status: Category A
Location: 55.953600, -3.180193

Perhaps no other city in the world embraced the Greek Revival as fervently as Edinburgh. Technical information about Grecian architecture had been filtering back to the Scottish capital since the middle of the eighteenth century, the most influential study being James Stuart and Nicholas Revett's *The Antiquities of Athens*. The ensuing years saw a proliferation of buildings constructed inspired by these classical forms, the style remaining popular here long after its influence had begun to wane elsewhere. Its impact on the cityscape is perhaps most obvious on Calton Hill, which is topped by the unfinished National Monument, intended to be a replica of the Parthenon. Just as funds for this ambitious project were exhausted, Hamilton's new home for the Royal High School was being completed.

Originally destined for a site in the New Town, the city was eventually compelled to use a difficult sloping site on the southern side of Calton Hill. Hamilton settled on a symmetrical plan, all in Greek Doric, with a central temple flanked by blank colonnades. These link to pavilions at the eastern and western extremities of the main building. Completing the composition of the southern elevation are projecting entrance portals and a pair of lesser 'temples' which follow the line of Regent Road. Decorative elements are minimal, this approach being carried over into the interiors. Exceptions to this can be found in the former assembly hall, which features a coffered ceiling and iron pillars with ornate capitals.

Ever since the Royal High School decamped to Barnton in 1968, successive administrations have failed to find a suitably prestigious role for this historic structure. Intended as the home of a devolved Scottish Assembly, conversion work was completed in anticipation of 'yes' votes topping the required 40 per cent, with the building being renamed New Parliament House. This threshold was not reached, and when a Scottish Parliament was finally created in 1998 New Parliament House was overlooked in favour of a site at Holyrood. Since then proposals have been put forward for conversion to a hotel and as a home for a national photography centre.

74 St Andrew's House
1930s purpose-built government office complex
www.scotland.gov.uk/About/Locations/St-Andrews-House-1
Year of completion: 1939
Architect: Thomas Tait
Listed Status: Category A
Location: 55.953600, -3.180193

Following decades of wrangling and controversy – which seem to be obligatory with any major construction project within the capital – this colossal new home for the Scottish Office was finally opened just as the

storm broke and war with Germany was declared in 1939. It occupies the site of Calton gaol, of which only the governor's house and a large curtain wall remain extant. No doubt aware of the weight of expectation attached to such a prestigious project, Tait's design was not in the vanguard of the modern movement, incorporating elements of classical, art deco and international modern design while integrating well with its Greek Revival neighbours. Northumbrian Darney stone was chosen to clad the steel frame, its pale colour very obvious in photographs taken soon after the building was completed. However airborne pollution soon tarnished the exterior, which wears a coat of soot and grime to this day.

Decorative elements are mostly restricted to the north façade, whose entrance is flanked by moderne lamp/flag standards. Main doors are cast in bronze and depict images of St Andrew. Above this an intricately carved royal coat of arms by Alexander Carrick. Perhaps most impressive are the series of sculptures which emerge from the top of the central pillars, executed by Sir William Reid Dick. From left to right they depict Architecture, Statecraft, Health, Agriculture, Fisheries and Education. All have acquired a noticeable green tinge thanks to an accumulation of moss.

75 The Nelson Monument
Castellated memorial tower used as signalling station
www.edinburghmuseums.org.uk
Year of completion: 1816
Architect: Robert Burn
Listed Status: Category A
Location: 55.954485, -3.182612

One of a collection of disparate structures all built within twenty years of each other, though the National Monument, 'Scotland's Disgrace', was never finished. Both this and the planned memorial to Lord Nelson were both funded partly through public subscription, which ran out before either could be completed. Thomas Bonnar eventually supervised the remaining work after the death of Burn in 1815.

What emerged was a six-storey ashlar tower, the uppermost level being set back to provide a viewing platform which affords panoramic views of Edinburgh and the Firth of Forth. At the tower's base a five-sided structure was intended to provide living quarters for the signals staff. A fortified style was deemed appropriate, with crenellations, blind quatrefoils and arrow slit windows. Conceived from the beginning as a functioning building as well as a monument, a later addition was to provide a crucial aid to navigation at sea. This was the time ball, installed on a mast-like arrangement at the top of the tower. When lowered this visual signal would provide ships' captains with an accurate time check, enabling them to set their chronometers and thereby calculate longitude.

76 Waterloo Place
Greek Revival style development at east end of Princes Street
Year of completion: 1819
Architect: Archibald Elliot
Listed status: Category A
Location: 55.953640, -3.188281

Extending Princes Street to the east created the opportunity for a ribbon of developments which were designed to reflect the growing civic pride and confidence of the era, with the burgeoning Greek Revival being

deemed the most appropriate architectural style to convey this. First to be constructed, at the west end of the new road and abutting Regent Bridge, Waterloo Place comprises four separate buildings that were intended for mixed residential and commercial occupation. Bookending the westernmost blocks are a pair of porticos on either side of the road which frame the view of Calton Hill, their uniformity only spoilt by some alterations to the ground floor at Number 1: at some point the sash windows have been replaced and the ashlar painted white.

For the most part the exteriors have survived largely as built, but the changing requirements of succeeding tenants have meant that most of the interior spaces have been remodelled. At the former Stamp Office only the façade remains, screening modern open-plan office space in the same manner as neighbouring Waverleygate. Next door, the old General Post Office (predecessor to Robert Matheson's building for the GPO of 1866) survived extensive fire damage in the 1950s and has now been converted into flats. Facing this is what was Edinburgh's first purpose-built hotel, recently restored to its original function after over a century of use as offices. Some internal features have been preserved, including an oak-panelled room now used as a conference suite.

77 Preserved section of tramway track
Remnant of defunct cable operated tramway system
Date of completion: 1888 (first cable operated services)
Location: 55.953503, -3.188197

Cross the road at the traffic lights outside *Waterloo Place* [76] and you may notice a small section of preserved cobbles, with three rails embedded within them. These are among the last visible remnants of an early mass transit system utilising a vehicle which is currently being reintroduced to the city – the tram. Early examples of these were pulled by teams of horses, which had the unenviable task of hauling the tramcars up the many steep Edinburgh gradients. Operation via cable was deemed to be more efficient, these being introduced in 1888. A stationary steam engine would wind in the cable, which was fed through an open conduit between the tracks and attached to the tram itself.

This method of propulsion was rather slow, which was just as well given the rudimentary braking systems employed. Another problem was the risk of the cable snapping, a regular occurrence which would leave the whole contraption stranded until repairs could be effected. By 1923 the entire network had been converted to run on electricity, before finally being replaced altogether by buses in 1956.

Trams will return to the city in 2014 upon completion of the highly controversial line running from Edinburgh Airport to York Place. Behind schedule, vastly over budget and reduced in scale, Edinburgh's new tramway system has been mired in controversy throughout its protracted construction process.

78 Monument to Duke of Wellington
Bronze equestrian statue
Year of completion: 1852
Sculptor: Sir John Steell
Listed status: Category A
Location: 55.953503, -3.188197

Situated outside the entrance to *General Register House* [79], the erection of this statue necessitated the realignment of that building's

front perimeter wall. Placed on top of a granite plinth, the duke gazes impassively towards North Bridge as his horse Copenhagen rears on its hind legs. This stance goes some way to refuting the misconception that military leaders depicted in this pose had died in battle. This portrayal of the Iron Duke has been spared the fate of a similar monument in Glasgow, which is usually to be found adorned with one or more traffic cones placed there by intrepid and irreverent locals.

In 2003 the statue was at the centre of some controversy when it was suggested (by an SNP election candidate) that it be removed and replaced with one of Robert Burns. Nothing became of this particular piece of political opportunism.

Steell was also responsible for the statue of Sir Walter Scott at the foot of his monument, a short distance away on Princes Street.

79 General Register House
Neoclassical record repository with domed central rotunda
www.gro-scotland.gov.uk
Year of completion: 1788
Architects: Robert and James Adam
Listed status: Category A
Location: 55.953787, -3.189314

By the middle of the eighteenth century it had become essential to create a permanent, secure home for the vast amount of governmental and legal records that had been steadily accumulating for centuries,

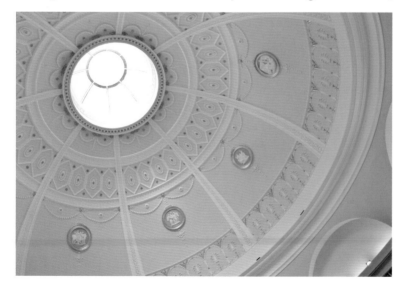

these often being stored in wholly unsuitable locations. A site was chosen opposite the old Theatre Royal (demolished to make way for *Waverley Gate* [115]), funding secured via the seized assets of displaced Jacobites and a renowned architectural family employed to execute the planned repository. It was the beginning of a long and often tortuous construction programme. A government grant of £12,000 was soon exhausted, with building work being suspended for six years. There were also financial disputes between the trustees and contractors. Robert Adam did not live to see his design fully realised, it being left to Robert Reid to complete the north side of the building in the 1820s.

Approaching from North Bridge allows time to appreciate the restrained neoclassical thirteen-bay south elevation, which is rusticated on the ground floor and features a central portico of four columns topped by Corinthian capitals. At each corner are projecting pavilions. It is possible for members of the public to access the archives that are still held here; those that do can take the opportunity to gaze up at the impressive top-lit central rotunda, with its intricate plasterwork completed by Thomas Clayton to Adam's designs.

80 The Balmoral Hotel
Imposing railway hotel with Renaissance detailing
www.thebalmoralhotel.com
Year of completion: 1902
Architect: William Hamilton Beattie
Listed status: Category B
Location: 55.952895, -3.189398

This behemoth looks down on the very point where the A1 completes its 396-mile journey from central London, at the junction with Princes Street, the A7 North Bridge and the A900. Though now long since accepted as a familiar element of the Edinburgh skyline, on completion the building was castigated for its intrusive scale (it towers over nearby *General Register House* [79]) and for bringing a tawdry commercial vulgarity to the south side of Princes Street. Fearful for the views of the castle from Calton Hill, the Cockburn Association successfully campaigned to have the clock tower reduced in height. Certainly the 'NB' as it was known (short for the North British Hotel) did not lack for ornamentation. Its architect had already designed the Jenners department store nearby which featured elaborately carved sandstone elevations. The North British is also a riot of classical Renaissance detailing: there are Ionic pilasters, stone balustrades, Dutch gables, turrets and corner domes. Soaring above all this, the clock tower wears an ornamental iron crown.

Built to serve Waverley station, the North British was also a potent symbol of the power and prestige of the railway company who commissioned it. No expense was spared in a bid to upstage their great rivals the Caledonian Railway, who were engaged in building their own hotel at the other end of Princes Street. Wealthy guests could be conveyed by lift directly from the station platform to the entrance hall, from where they could easily reach the dining areas including the opulent Palm Court. Nowadays the direct link from station to hotel has been severed, the basement entrance replaced by a swimming pool. One tradition has been maintained however: the hotel clock is still set a few minutes fast in an attempt to ensure travellers catch their trains on time.

A thorough restoration of the building was completed in 1991, with cleaning of the pollution-stained stonework undertaken by Stonewest Cox.

Other Landmarks

London

81. One London Wall (City): Glass and steel office block incorporating livery hall.
82. Crescent House, Golden Lane Estate (City): Mixed residential/commercial block by Chamberlin, Powell & Bon.
83. Everyman Screen on the Green (Islington): Very early purpose-built cinema.
84. Almeida Theatre (Islington): Theatre converted from Islington Literary and Scientific Institute. Stripped classical façade.
85. Islington Town Hall: Mixed classical designs housing council chambers and events venue.
86. St Gabriel (Archway): Modernist 1960s Roman Catholic church.
87. Jackson's Lane Community Centre (Highgate): Conversion of Methodist church. Gothic revival style.
88. Hendon Hall Court (Hendon): Concrete brutalist flats by the Owen Luder Partnership.
89. WLMG Nissan (Mill Hill): 1960s car dealership and garage sandwiched between M1 and A1. Concrete brutalist aesthetic.
90. Former RNLI Depot (Borehamwood): Storage facility in Streamline Moderne style.

Southern England and the Midlands

91. Black Cat roundabout (Bedfordshire): Metal feline sculpture commemorating former garage.
92. Biggleswade Wind Farm (Bedfordshire): Group of ten 2 MW wind turbines.
93. The Norman Cross Memorial (Cambridgeshire): Bronze eagle sculpture commemorating first purpose-built prisoner-of-war camp in Britain.
94. Norman Cross Motel (Cambridgeshire): Rare surviving 1960s motel. Circular plan.
95. Wansford Station (Cambridgeshire): Station building and signal box on preserved Nene Valley Railway.
96. Gaydon Hangar, RAF Wittering (Cambridgeshire): Aircraft hangar designed to house Cold War-era nuclear bombers.
97. Wothorpe Towers (Cambridgeshire): Remains of seventeenth-century lodge built for the Earl of Exeter.
98. St Peter, Tickencote (Rutland): Late eighteenth-century reconstruction of Norman church.

99. Tickencote OK Diner (Rutland): Pastiche of 1950s US roadside diner clad in stainless steel.
100. Ram Jam Inn at Stretton (Leicestershire): Coaching-era inn said to have been frequented by Dick Turpin. Closed as of 2014.
101. Newlink Business Park (Nottinghamshire): Pair of vast steel-framed distribution warehouses.
102. St Wilfred, South Muskham (Nottinghamshire): Medieval parish church.
103. St Nicholas, Tuxford (Nottinghamshire): Large parish church serving former market town.

The North

104. Cusworth Hall, Doncaster (South Yorkshire): Mid-eighteenth-century Palladian country house.
105. Holmfield Interchange bridge piers (West Yorkshire): 'Wineglass' shaped concrete piers supporting motorway flyovers.
106. Watch Office at Former RAF Catterick (North Yorkshire): Very early example of aircraft control tower.
107. Intu Metrocentre (Tyne & Wear) Vast out-of-town retail and leisure centre.
108. Gateshead Mazda car storage tower (Tyne & Wear): Glass car display tower incorporating hydraulic lifts.

Scotland

109. Houndwood Church (Berwickshire): Former parish church built in classical style.
110. Oxwell Mains windpump (East Lothian): Masonry shell of former drainage windpump.
111. New Restalrig Parish Church (Willowbrae Road, Edinburgh): Gothic revival parish church.
112. Police Box (Comely Green Place, Edinburgh): Dilapidated 1930s former police box. Grecian-influenced design.
113. Artisan Bar (London Road, Edinburgh): Well-preserved end-of-terrace public house. Completed 1871.
114. Regent Terrace (Edinburgh): Terrace of residential town houses with classical detailing.
115. Waverleygate (Waterloo Place, Edinburgh): Multi-let office development behind retained Italianate façade.

Bibliography

Cottam, David, *Sir Owen Williams 1890–1969* (The Architectural Association, 1986).

Daiches, David, *Edinburgh* (Hamish Hamilton, 1978).

Denison, Edward and Ian Stewart, *How to Read Bridges* (Herbert Press, 2012).

Dixon, Roger and Stefan Muthesius, *Victorian Architecture* (Thames & Hudson, 1978).

Eyles, Allen, *Gaumont British Cinemas* (BFI Publishing, 1996).

Gifford, John, Colin McWilliam and David Walker, *The Buildings of Scotland: Edinburgh* (Penguin, 1984).

Glendinning, Miles and Aonghus MacKechnie, *Scottish Architecture* (Thames & Hudson, 2004).

Harbison, Robert, *The Shell Guide to English Parish Churches* (André Deutsch Limited, 1992).

Harwood, Elain, *England: A Guide to Post-War Listed Buildings* (English Heritage, 2003).

Hatherley, Owen, *A Guide to the New Ruins of Great Britain* (Verso, 2010).

Headley, Gwyn and Wim Meulenkamp, *Follies: A National Trust Guide* (Jonathan Cape Ltd, 1986).

Hutchinson, John, E. H. Gombrich, Lela B. Njatin and W. J. T. Mitchell, *Antony Gormley* (Phaidon, 2000).

Ison, Leonora and Walter, *English Church Architecture Through the Ages* (Arthur Barker, 1972).

Jenkins, Simon, *England's Thousand Best Churches* (The Penguin Press, 1999).

Jones, Edward and Christopher Woodward, *A Guide to the Architecture of London* (Weidenfeld & Nicolson, 2009).

Kadish, Sharman, *The Synagogues of Britain and Ireland* (Yale University Press, 2011).

Kirby, Sue and Richard Busby, *Hatfield: A Pictorial History* (Phillimore & Co. Ltd, 1985).

Mckean, Charles, *Edinburgh: An Illustrated Architectural Guide* (RIAS Publications).

Moore, Colin, *Windmills, A New History* (The History Press, 2010).

Moran, Joe, *On Roads: A Hidden History* (Profile Books, 2009).

Morrison, Kathryn A. and John Minnis, *Carscapes: The Motor Car, Architecture and Landscape in England* (Yale University Press, 2012).

Mullay, Sandy, *The Edinburgh Encyclopedia* (Mainstream Publishing, 1996).

Nicholson, Jon, *A1: Portrait of a Road* (Harper Collins, 2000).

Pevsner, Nikolaus, *The Buildings of England: Nottinghamshire* (Penguin, 1951).

Pevsner, Nikolaus and Ian Richmond, *The Buildings of England: Northumberland* (Yale University Press, 2002).

Pevsner, Nikolaus and John Harris, *The Buildings of England: Lincolnshire* (Yale University Press, 2002).

Saint, Andrew, *Richard Norman Shaw* (Yale University Press, 1976).

Sharp, C. Martin, *DH: A History of de Havilland* (Airlife Publishing, 1982).

Smith, Edwin, Olive Cook and Graham Hutton, *English Parish Churches* (Thames & Hudson, 1976).

Wright, Herbert, *London High: A Guide to the Past, Present and Future of London's Skyscrapers* (Frances Lincoln, 2006).

Websites

www.bbc.co.uk

www.rcahms.gov.uk

www.bwtas.blogspot.com

www.pylons.org

www.british-history.ac.uk

www.english-heritage.org.uk

www.nls.uk (National Library of Scotland)

www.historicengland.org.uk

www.scottisharchitects.org.uk

www.cbrd.co.uk

www.milestonesociety.co.uk

Booklets and Papers

East, John, *Civic Plunge Revisited* (Twentieth Century Society, 2012).

Gould, M. H. and D. J. Cleland, *Development of Design Form of Reinforced Concrete Water Towers*.

Hyde, Bert, *Stotfold Pubs and Publicans* (2002).

Metheringham, S. and J. Townshend, *A1 River Tyne Bridge, UK – Design and Construction* (Institute of Civil Engineers, 2005).

Sanderson, Margaret H. B., *'A Proper Repository', The Building of the General Register House* (Scottish Record Office, 1992).

Stuart-Mogg, David, *The Story of Wansford*.

Szynalska, Karolina, *Sam Scorer: A Lesser Known Architect of the Twentieth Century* (September 2010).

Taylor, Simon, Matthew Whitfield and Susie Barson, *The English Public Library 1850–1939* (English Heritage, 2014).

The Architects' Journal Information Library, 23 September 1970 (Meadowbank Sports Centre).